Statistics
at the Bench

A Step-by-Step Handbook for Biologists

ALSO FROM COLD SPRING HARBOR LABORATORY PRESS

RELATED TITLES

Genetics of Complex Human Diseases
Bioinformatics: Sequence and Genome Analysis, Second Edition

RELATED HANDBOOKS

At the Bench: A Laboratory Navigator, Updated Edition
At the Helm: A Laboratory Navigator
Binding and Kinetics for Molecular Biologists
Experimental Design for Biologists
Lab Math: A Handbook of Measurements, Calculations, and Other Quantitative
*   Skills for Use at the Bench*

Statistics at the Bench

A Step-by-Step Handbook for Biologists

M. Bremer

Department of Mathematics
San Jose State University

R.W. Doerge

Department of Statistics
Department of Agronomy
Purdue University

COLD SPRING HARBOR LABORATORY PRESS
Cold Spring Harbor, New York · www.cshlpress.com

Statistics at the Bench
A Step-by-Step Handbook for Biologists

Publisher	John Inglis
Acquisition Editor	Alexander Gann
Development Director	Jan Argentine
Project Coordinator	Inez Saliano
Production Manager	Denise Weiss
Production Editor	Rena Steuer
Marketing Manager	Ingrid Benirschke
Sales Manager	Elizabeth Powers
Cover Designer	Ed Atkeson

Front cover illustration was created by Jim Duffy.

Library of Congress Cataloging-in-Publication Data

Bremer, M. (Martina)
    Statistics at the bench : a step-by-step handbook for biologists / M.
Bremer and R.W. Doerge.
        p. cm.
    Includes indexes.
    ISBN 978-0-87969-857-7 (hardcover : alk. paper)
    1. Biometry. I. Doerge, R. W. (Rebecca W.) II. Title.

QH323.5.B744 2009
570.1'5195--dc22

                                                                2009023376

10 9 8 7 6 5 4 3

All Cold Spring Harbor Laboratory Press publications may be ordered directly from Cold Spring Harbor Laboratory Press, 500 Sunnyside Blvd., Woodbury, New York 11797-2924. Phone: 1-800-843-4388 in Continental U.S. and Canada. All other locations: (516) 422-4100. FAX: (516) 422-4097. E-mail: cshpress@cshl.edu. For a complete catalog of all Cold Spring Harbor Laboratory Press publications, visit our website at http://www.cshlpress.org/.

Contents

Acknowledgments

We would like to thank the Department of Statistics at Purdue University and ADG for initiating the serendipitous circumstances that brought us together. We are grateful to our families and friends for their support during the pursuit of this project. Our thanks also go to Andrea Rau and Ben Hecht for their insightful comments and suggestions that helped to improve this Manual.

MARTINA BREMER
REBECCA W. DOERGE

1 Introduction

Biology is becoming increasingly computational. New technologies are producing massive amounts of data, particularly in molecular biology, and are opening entirely new avenues of research. However, with these new technologies come new challenges and needs. Large amounts of quantitative information need to be organized, displayed, and understood. Even for biologists who are not directly involved in collecting data, thinking quantitatively and interpreting quantitative results, both in research and in the literature, are becoming important aspects of daily scientific training and work.

Large-scale sequencing projects are good examples of how science enabled by technology has given rise to massive amounts of data (e.g., genome and protein sequences) that need to be summarized, investigated, and quantitatively assembled. Case in point: the simplest software program used for assessing sequencing data is BLAST (Basic Local Alignment Search Tool), in which every search conducted by the program returns statistical summaries that describe how well search results "fit" a query. Understanding what these results are actually telling us is essential to making well-informed scientific choices.

Having an appreciation for the computer results provided by software applications or quantitative results from the literature requires a basic understanding of statistical thinking and statistical procedures. Likewise, quantitative reasoning about the design of an experiment and data evaluation are as important as formulating and answering biological questions based on challenging and complex experimental situations. With this in mind, this Manual is intended as a resource that provides an overview of terminology and concepts that occur frequently in quantitative thinking and in the statistical analysis of biological data. This Manual also explains, in the simplest of terms, the underlying principles of the more complex analyses (e.g., microarray experiments).

If you are looking for a one- or two-semester textbook on statistical procedures, you have the wrong book in hand—this Manual is not intended as a textbook. Instead, this is a "bench-side" Manual that is designed and intended to be utilized by people who are in need of a quick refresher or a big-picture overview of a statistical procedure. Maybe it has been a long time (if ever) since you last took a course in mathematics or statistics, or even considered thinking quantitatively about your data. Maybe the amount of quantitative training you received is lacking or less than desirable. This Manual is aimed at you! The comprehensive index allows you to quickly access the information you need to think about your data and interpret your results effectively.

As you will see and read in this Manual, new concepts are illustrated by simple examples selected from different biological applications. Many of the examples that we use are accompanied by detailed Excel commands. We chose Excel because it is one of the most commonly available software applications and is widely used by biologists today. This said, we recognize the limitations of Excel and we acknowledge that it is not an advanced statistical software program. For applications that exceed the scope of this text (and Excel), there are other statistics programs that may be more appropriate (for a discussion of available programs, read section 2.7).

> **In Excel:** Complete Excel examples can be found in boxes throughout the text. Some require the use of an "Add-in" package. To access the package, open Excel and click on TOOLS → ADD-INS... and checkmark the ANALYSIS TOOLPAK and the ANALYSIS TOOLPAK-VBA.

Today, computationally trained biologists are gaining a competitive edge in science. Not only can they read the literature, they can also think critically about it and engage in meaningful scientific conversations that are rich with quantitative reasoning. Our purpose in providing this Manual is to assist biologists in becoming fluent and comfortable in the language of quantitative reasoning and to facilitate open and informed communication between the biological and the quantitative sciences.

2 Common Pitfalls

2.1 Examples of Common Mistakes

The interesting thing about Statistics as a discipline is that it allows anyone to summarize and analyze data. After loading data into one's favorite statistical software package (or at least the one that is available), 9 times out of 10 just about anyone can produce some sort of numbers or answers. Yes, anyone can do a "statistical analysis" of their data, which is perilous if analyses are not done correctly. With this in mind, we list a number of common, but unintentional, mistakes that occur either when designing an experiment or analyzing the data from an experiment.

- No clear statement of the scientific question(s).
- Collecting the wrong data.
- Lack of randomization.
- Ignoring the importance of biological replication.
- Making things more complicated than they need to be.
- Ignoring obvious components of the experiment.
- Using the wrong statistical procedure.
- Ignoring the assumptions of the model.
- Altering the experimental question, after the fact, to fit the results.
- Losing the big picture of the experiment when thinking about the results from the analyses.

Many of these mistakes result in frustrations that can be avoided simply by being aware of the importance of good statistical design and analysis.

2.2 Defining Your Question

"Finding the right question is often more important than finding the right answer" (Tukey 1980). It seems like an obvious point, but it is important to know what question(s) you want to ask, and then devise a plan or sampling design to collect the data that will help answer the question(s). One common mistake is that the wrong data are often collected to address the right questions. Thinking about your question and what data need to be collected before performing a study will save time, money, and frustration. The old adage, consult a statistician prior to designing your experiment, actually is true.

Questions to ask yourself:

- What are my questions?
- What data do I need to collect? What data am I able to collect?
- How do I measure the data accurately?
- What external forces are acting on my biological system? Can I measure them?
- How does variation affect my data and the results?
- How will I report my results in a comprehensive manner?

2.3   Working with and Talking to a Statistician

Vocabulary is the largest divide between the biological sciences and the mathematical sciences. The second largest issue is communication. Statisticians and biologists fundamentally communicate very differently. Coupling vocabulary and communication issues into any human interaction usually results in frustrations.

Statements from statisticians that biologists are tired of hearing. There are a number of statements that frustrate biologists, and frankly, they are tired of hearing them. Here are some examples:

- Consult a statistician prior to designing your experiment.
- What is your null hypothesis?
- Your sample size is too small.
- You did not replicate.
- You did not randomize.
- You collected the wrong data.

Equally, there are a number of fundamentally huge issues, at least from a statistician's point of view, that biologists tend to sweep under the carpet as unimportant and high-headed mathematical mumbo-jumbo.

Statements from biologists that statisticians are tired of hearing:

- If you need statistics, your result is not real.
- I did not replicate because the result might change (or, I don't have enough money).
- Is there a difference between a technical replicate and a biological replicate?
- I got the answer from my favorite software package; I have no idea what I did.
- I just want a small p-value.

The majority of biologists attempt to design their own experiments and analyze their own experimental data, and only if a problem arises do they consult a statistician. This consultation usually takes place after the data are collected, when the frustrations and anxiety are high. As such, the best way to approach a consultation with a statistician is to be prepared. If the data are in hand, before the meeting send a brief document to the statistician that includes the questions you are addressing, a description of the data that were collected (include pictures), and the conditions under which the data were collected. If the initial consultation is prior to actually performing the experiment, the statistician may work with you to create such a document, which will then become your experimental plan and design.

2.4   Exploratory versus Inferential Statistics

Much of biology can be viewed as observational. Data are collected, summary statistics calculated, and pictures drawn. From this exploratory information, patterns are often observed, and it is quite easy to start drawing conclusions prior to actually controlling for the experimental conditions and sources of variation.

- A common mistake is to explore data without incorporating the experimental design, which in turn may result in observational patterns that are due to external sources (e.g., dye bias in a microarray experiment that implies differential expression of a gene). Conclusions drawn from such experiments may be misleading if the external sources of variation have a significant effect on the outcome.

Inferential statistics is based on statistical theory, and although it may seem complicated, it is rather simple. A mathematical model describes the relationship between the variable(s) of interest (i.e., the dependent variable) and the other variables (i.e., the independent variables) of the experiment. Using this relationship, questions can be asked of the data. The simplest models are linear in nature, with the most basic being an equation for a straight line ($y = mx + b$). Answers are gained by comparing a quantity, or test statistic, calculated from the experimental data to the situation where there is no relationship (or a random relationship) between the variable of interest and the independent variable(s). If the quantity calculated from the experimental data is more likely than the random representation, then the relationship between the dependent variable and the independent variable(s) is significant in nature.

- Some of the most common pitfalls of inferential statistics are to make the initial relationship between the dependent variable and independent variable too complicated, to ask an incorrect question (i.e., test the wrong quantity), and to ignore the assumptions that are required for the data prior to analysis.

2.5 Different Sources of Variation

Variation is not necessarily a bad thing, if it can be understood and dealt with properly. Every variable in an experiment that is measured quantitatively will have variability both because no single measuring device is perfect and because many external forces may be acting on the experimental system under investigation.

- A common misconception among biologists is that variation is bad and that if the experiment is repeated, or if the measurement is taken again, the quantity should be exactly the same.

  - For example, two technicians scoring quantitative disease resistance will most likely obtain two different measurements. Certainly, the experimental unit has not changed,

2.6 Model Assumptions Are Important

but the variation due to the technicians may cause the
measurements to be different. The good news is that we
can control for this source of variation (i.e., variation due
to technician) in the statistical model and analysis.

- Another common mistake is to disregard sources of technical
  variation and to only concentrate on the biological variation. In
  doing so, technical variation is confounded with, or included in,
  the variation that is observed biologically, making it impossible
  to obtain an accurate assessment of biological variation.

Often, biological variation is considered unimportant, and as such
it is only deemed necessary to measure the experimental unit once.
Because experimental units, especially in biology, interact differently
with the technology, environment, and other influences, it is important to fully understand how much biological variation there is and
how it differs from technical variation. It is worth remembering that
the role of statistics is to partition variation into its assigned components so that the relationship between the (dependent) variable of
interest and the sources of variation is well-defined.

- If important variables are left out of both the model and the
  analysis, then non-random patterns in the remaining variation
  (i.e., the residuals) may actually influence the results.

2.6   The Importance of Checking Assumptions and the Ramifications of Ignoring the Obvious

Anyone who has attempted to perform a statistical analysis may
realize there are assumptions about the data that should be satisfied
prior to performing the analysis. The most frequent assumption that
is required of the data is that they are normally distributed. In fact,
most statistical analysis procedures are based upon the assumption
that the data are normally distributed. Ignoring the assumption of
normality usually leads to incorrect conclusions. The most common
mistake when performing a statistical analysis that requires data to
be normally distributed is to proceed with the analysis and ignore
the data's non-normality.

As with almost anything in life, if you ignore the problem (i.e., violation of assumptions), it may only get bigger. The ramifications of

ignoring the basic assumption of how the data and/or residuals be-
have, or are distributed, can result in misleading and incorrect results
and conclusions. Here are some of the most common assumptions
that are ignored:

- Data are normally distributed.
- All members of the population have the same variance (i.e.,
  common variance assumption).
- Once the data are fit to a model, residuals are normally dis-
  tributed.
- The residuals are random and pattern-free.

Unfortunately, the popularity of point-and-click statistical analysis
software has only compounded the problem of ignoring assumptions
since it is possible to jump into the middle of an analysis without
considering assumptions about the data, the general context of the
experiment, or basic questions upon which the study was formulated.

2.7 Statistical Software Packages

Personal computing has certainly made our lives more efficient and
our work lives more informed. Statistics, like many disciplines, has
greatly benefited from the implementation of many seemingly complex analyses into simple point-and-click software packages with
user-friendly pull-down menus full of summary statistics, analyses,
and graphical options. Some of the more popular statistical packages
are:

- SAS
- R
- JMP
- Minitab
- SPSS
- Mathematica

Although useful to statisticians, most of these computer packages are
frustrating for biologists because it may be difficult to load the data
into the program, challenging to understand the code required (since
not all packages have pull-down menus), and frustrating to interpret
the output in the context of the problem under investigation. Some of

the more common pitfalls when using unfamilar statistical software include the following:

- Data format does not match requirements of software package.
- It is unclear what to do with missing data, or how to code them.
- The order in which the exploratory investigations, graphics, and analyses should be performed is not straightforward.
- The output of the analysis does not make sense.
- The graphics are difficult to use.

With these frustrations in mind, and because we desire to assist biologists with their science at the bench, we are presenting all examples in this reference guide using Microsoft Excel. While Excel is not a statistical software package, it does have the capability to enable the user to explore and ask intelligent questions of her data, all easily accomplished on a laptop or desktop computer—at the bench.

3   Descriptive Statistics

3.1   Definitions

VARIABLE: A variable is an observed category (label) or quantity (number) in an experiment. Depending on whether the variable takes on label or number values, it is called a *categorical* or *quantitative* variable.

CATEGORICAL VARIABLE: A categorical variable takes on label values (e.g., red, green, blue). It can also take on number values if the numbers are used to encode for certain conditions (e.g., male $= 1$, female $= 2$). However, in this case the numbers should not be understood as numerical measurements (i.e., male $\neq \frac{1}{2}$ female).

QUANTITATIVE VARIABLE: A variable that counts the occurrence of an event or measures a numerical quantity. This kind of variable only takes on number values. If the variable is used to measure a count, it is called *discrete*. If the variable is used to measure a quantity which may take on any value in an interval (e.g., 0 to 1), it is called *continuous*.

ORDINAL VARIABLES: If the possible values of a variable possess a natural logical order (e.g., $1 < 2 < 3$ or strongly disagree $<$ disagree $<$ neutral $<$ agree $<$ strongly agree), the variable is called ordinal. In general, quantitative variables that measure something are always ordinal. Categorical variables may or may not be ordinal. Quantitative ordinal variables are called QUALITATIVE variables.

Example 3.1

To understand the different types of variables involved in biological experiments, consider the following examples:

(a) In Gregor Mendel's experiment (Mendel 1865) he recorded the color and shape of pea seeds. The possible values for color were yellow and green, and the possible values for shape were smooth and wrinkled, so both of these variables are categorical. Since there is no logical order in these values (e.g., green $\not>$ yellow and yellow $\not>$ green), neither variable is ordinal.

(b) If bacteria are grown on a Petri dish, then the number of bacteria that are visible in a randomly chosen 10^{-4} in^2 microscope cross-section is a quantitative random variable. The possible values are $0, 1, 2, 3, \ldots$. Since this variable is quantitative, it is automatically ordinal $(0 < 1 < 2 \cdots)$.

(c) In spectrophotometry, light is passed through a liquid and the amount of light absorption (in %) can be used to deduce the concentration of a substance in the liquid. Absorption in this case is a continuous variable that takes on values in the interval $[0, 1]$ or $[0, 100]$, depending on the scale that is used.

RANDOM VARIABLE: A random variable is a placeholder for the outcome of an experiment that is subject to chance. The value that the variable will take on is unknown before the experiment is conducted. However, the *distribution* of the random variable, which assigns probabilities to specific outcomes, may be known.

OBSERVATION: After an experiment has been conducted and is complete, the values that a random variable has taken on are recorded. These data are referred to as the observations in the experiment.

Example 3.2

Gene expression is a continuous random variable since it takes on values in an interval. Even though fold changes are sometimes reported as integers, intermediate values are possible and can be determined with more precision in the measurements. Thus, "expression" of a gene is a random variable. A description such as "gene XYZ underwent a 2.37-fold change between treatment and control sample" is an observation on the random variable.

When a random variable such as gene expression or height of a plant is observed repeatedly on several subjects (different genes, plants, etc.), it is often convenient to graphically display the results. For quantitative variables, the most commonly used graphical display methods are histograms, dot plots, scatter plots, and box plots, described in Section 3.3.

3.2 Numerical Ways to Describe Data

When several observations of the same type are obtained (e.g., gene expression of many genes in a microarray experiment) then it is often desirable to report the findings in a summarized way rather than list all individual measurements. This can be done in different ways. We next describe the most common numerical measures that summarize categorical and quantitative data.

3.2.1 Categorical Data

Categorical data are most commonly summarized in tables. The tables contain the labels or possible categories and COUNTS of how often these labels were observed in the experiment. If two categorical variables are observed on the same individuals, then the data can be summarized in the form of a two-dimensional CONTINGENCY TABLE.

Example 3.3

To test the effectiveness of two potential yellow fever vaccines A and B, laboratory mice are vaccinated with vaccine type A, or vaccine type B. Some mice are left unvaccinated to function as a control. All mice are infected with the yellow fever virus and, after an appropriate incubation period, live and dead mice are counted. Thus, data are collected on two categorical variables per mouse. One variable describes the type of vaccine the mouse received (A, B, or none), and the other variable states whether the mouse is alive or dead. The experimental results in the form of a contingency table look like this:

	A	B	None
Live	7	5	3
Dead	3	7	12

For example, ten mice were vaccinated with the type A virus, and of those seven survived.

3.2.2 Quantitative Data

MEAN: The average of all observations:

$$\bar{x} = \frac{1}{n} \sum_{i=1}^{n} x_i,$$

where the observations are denoted $x_1, x_2, \ldots, x_n$, and n is the number of observations (sample size). Other commonly used terms for the mean are EXPECTED VALUE and AVERAGE.

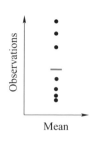

In Excel: To compute a mean, click on any empty cell, type "=AVERAGE()" and highlight the observations for which you want to compute the mean.

MEDIAN: The middle observation if the number of observations n is odd. If n is even, then the median is the average of the two middle observations. Half of the observations are always larger than the median and the other half are smaller.

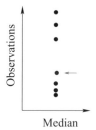

In Excel: To compute a median, click on any empty cell, type "=MEDIAN()," and highlight the observations for which you want to compute the median.

PERCENTILES: Similar to the median, the pth percentile of the observations is the observation value such that $p\%$ of the observations are smaller than it. Consequently, the median can also be thought of as the 50th percentile. The 25th and 75th percentiles are sometimes referred to as the first and third QUARTILES, respectively.

In Excel: To compute a percentile, click on any empty cell and type "=PERCENTILE(DATA, p)." Here DATA stands for your array of observations (highlight them) and p is the percentile that you want to compute as a number between 0 and 1. For example, an alternative for the median is to type "=PERCENTILE(DATA, 0.5)." The first and third quartiles can be obtained with the commands "=PERCENTILE(DATA, 0.25)" and "=PERCENTILE(DATA, 0.75)," respectively.

VARIANCE: The variance is the average squared
distance (deviation) of observations from the mean.

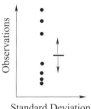

$$\text{Variance} \; = \; \frac{1}{n-1} \sum_{i=1}^{n} (x_i - \bar{x})^2 \,,$$

where $x_1, \ldots, x_n$ denote the observations and n
is the sample size. Variance is used as a measure
of variation in the observations and describes the
spread of the distribution.

STANDARD DEVIATION is simply the square root of the variance. Unlike the variance, which has nonsensical units, however, its units are
the same as the original observation measurements. Commonly used
symbols for standard deviation (variance) are s, σ (s^2, σ^2). Standard
deviation should not be confused with standard error (Section 3.6).

In Excel: To compute a variance, click on any empty cell, type
"=VAR()," and highlight the data. To compute a standard deviation,
click on any empty cell, type "=STDEV()," and highlight the data for
which you want to compute the standard deviation.

RANGE: This is the distance between the largest and
smallest observation. The range can be used as a crude
measure of spread in the data, but it is more susceptible than the variance (standard deviation) to mis-
represent the true variation in the data if there are
uncharacteristically large or small observations among
the data points.

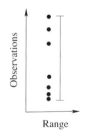

In Excel: To compute a range in Excel, determine the largest and
smallest data points, take their difference by typing "=MAX(DATA) −
MIN(DATA)" into any empty cell, and highlight your data.

INTERQUARTILE RANGE: Another measure for variability in the data
which is less dependent on extreme (big or small) data values is the
interquartile range (IQR). It is the distance between the third and
first quartiles of the data. This unit is sometimes used to determine
whether or not a data point may be considered an OUTLIER.

In Excel: To compute the interquartile range in Excel, determine the third and first quartiles of your data and compute their difference. Click on any empty cell and type "=PERCENTILE(DATA, 0.75) − PERCENTILE(DATA, 0.25)."

3.2.3 Determining Outliers

An outlier is a data point whose value is very different from the majority of the rest of the data. Outliers can be caused by errors during the measurement process or during the recording of data. Outliers may also be measurements that differ from the majority of the data points for legitimate reasons (e.g., one patient with a rare extreme reaction to a treatment). The decision as to whether or not an observation is an outlier is a subjective one.

A statistical rule of thumb to decide whether or not an observation may be considered an outlier uses the IQR of the data. Using all data points (including possibly suspected outliers), compute the first and third quartiles Q_1 and Q_3 as well as the interquartile range IQR $= Q_3 - Q_1$ for the data. An observation can be considered an outlier if it is either larger than $Q_3 + 1.5 \times$ IQR or smaller than $Q_1 - 1.5 \times$ IQR.

If an observation is suspected to be an outlier, double-check the recording of the observation to rule out typographical errors. If the measurement cannot be repeated, a statistical analysis should be performed with and without the outlier. If the data point is omitted in the subsequent analysis, then this should be accompanied by an explanation of why this value is considered to be an outlier and what may have caused it to be so different from the bulk of the observations.

Example 3.4

In 1902, the iron content of various foods was measured by G. von Bunge (1902). In this experiment, spinach was determined to contain 35 mg of iron per 100 g of spinach. When this value was used by other scientists later on, an error was made. The scientists used the value measured by von Bunge, but failed to notice that it was attributed to dried spinach rather than raw leaves. This error gave rise to a campaign for the health benefits of spinach as well as to a popular comic figure.

The table below lists iron content (in mg) in 100 g of particular foods as reported by the USDA in the national nutrient database for standard reference (U.S. Department of Agriculture 2005).

Food	Iron per 100 g (in mg)
Beef, cooked	6.16
Sunflower seeds, roasted, salted	3.81
Chocolate, semisweet	3.13
Tomato paste, canned	2.98
Kidney beans, boiled	2.94
Spinach, raw	2.70
Brussel sprouts, cooked	1.20
Soy milk	1.10
Lettuce, raw	1.00
Broccoli, raw	0.91
Cabbage, red, raw	0.80
Raspberries, raw	0.69
Strawberry, raw	0.42
Potato, baked	0.35

Food (von Bunge 1902)	Iron per 100 g (in mg)
Spinach, dried	35.00

If these values were combined, would the high iron content of dried spinach be considered a statistical outlier? What about the value for beef? Computing the quartiles for the above 15 data points yields $Q_1 = 0.855$ and $Q_3 = 3.055$ and hence an interquartile range of

$3.055 - 0.855 = 2.2$. Since the value 35 is larger than $Q_3 + 1.5 \times$ IQR $= 3.055 + 1.5 \times 2.2 = 6.355$, the dried spinach value would be considered an outlier in this data set. This makes sense, as all other values are taken from fresh or cooked foods and the dried spinach is fundamentally different from all other foods listed. The iron value for beef (6.16 mg) is smaller than $Q_3 + 1.5 \times IQR$ and hence would not be considered an outlier.

3.2.4   How to Choose a Descriptive Measure

The mean and median are numerical measures that are used to describe the center of a distribution or to find a "typical" value for a given set of observations. If there are some atypical values (outliers) among the observations, then the median is a more reliable measure of center than the mean. On the other hand, if there are no outliers and especially if the number of observations is large, then the mean is the preferred measure of center. If the data are to be used to answer a specific question, then working with means rather than medians will make subsequent statistical analysis much more straightforward.

Variance, standard deviation, range, and interquartile range are all measures that can be used to describe the spread or variability in the observations. If the variability in the data is small, then this means that the observations are all grouped closely together. If, on the other hand the observations cover a wide range of values, then the variation measure will be large. Similar to the case of center measure, the variance and standard deviation are better suited in cases where there are no extreme outliers among the observations. The range is very susceptible to outliers as it is based entirely on the largest and smallest observation values.

Data characteristic	Statistical measure	When to use
Center	mean	no outliers, large sample
	median	possible outliers
Variability	standard deviation	no outliers, large sample
	interquartile range	possible outliers
	range	use with caution

3.3   Graphical Methods to Display Data

BAR PLOT: For categorical data, a common graphical display method is a bar plot. The numbers of observations that fall into each category are counted and displayed as bars. The lengths of the bars represents the frequency for each category. Since the values of a categorical variable may not be ordinal, the order of the bars (each labeled by the category it represents) can be altered without changing the meaning of the plot.

Example 3.5

Suppose the variable recorded is flower color of a plant. In an experiment, 206 progeny of a certain parental cross of a flowering plant were obtained and categorized by their flower color. The results may be listed in table form. Figure 1 displays the data in the form of two bar plots.

Color	Count
red	94
white	37
pink	75

PIE CHART: When displaying categorical data, an alternative to the bar graph is a pie chart (compare Figs. 1 and 2), in which the frequencies (number of observations of a particular category) are expressed as relative frequencies. To compute relative frequencies (% of total observations), determine the total number of observations and divide each frequency by that number. The categories are then represented by "slices of pie" of varying thicknesses rather than by bars. Together, the slices add up to the "whole pie" (Fig. 2, circle) and represent the total number of observations.

> **In Excel:** Write the category labels and corresponding frequencies in table form. The order in which the categories appear in the table will determine the order in which they appear in the graph. Highlight the table and click INSERT → CHART... and select the appropriate chart type.

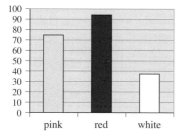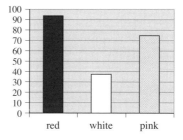

FIGURE 1. Observed flower color of 206 plants plotted as a bar plot. The bars are labeled by the colors they represent and the frequencies are expressed by the lengths of the bars. The order of the categories is arbitrary.

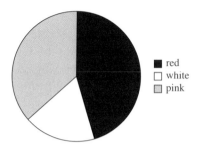

Figure 2. Pie chart of the flower color data from Example 3.5.

Histogram: In a histogram, the range of observations is divided into subcategories (most often of equal size). The frequencies of observations (i.e., number of observations that fall into a certain subcategory) are plotted as a bar on the y-axis. The bin width and number of bins that should be used in a histogram depend on the data. Histograms may not be appropriate if the number of observations is small. The larger the number of observations, the narrower the bins can be chosen while still accurately portraying the data. Well-constructed histograms allow a quick overview of the data to check for center, symmetry, outliers, and general shape.

Example 3.6

In 1846, William A. Guy studied the duration of life among the male English gentry (Guy 1846). He recorded the life spans of 2455 adults (ages 21 and up) from pedigrees of country families and from mural tablets. His results are displayed in Figure 3 in the form of a histogram. Since the number of observations (2455) in this example is very large, the histogram bin width can be relatively small (e.g., 5 years or less) to obtain a detailed impression of the results. If the bin width is larger (e.g., 10 years), the presentation of the same data will have less detail in its visual distribution. The size of the bin width should be determined by the level of detail needed in the study's conclusions.

In Excel: To create a histogram, write your observations into a column (in any order). Click Tools → Data Analysis → Histogram → OK. For the input range, highlight your data column. You can specify the subcategories by specifying values to use for the bins. In Example 3.6, the bin boundaries were chosen to be $20, 30, \ldots, 110$ for the left graph

continued

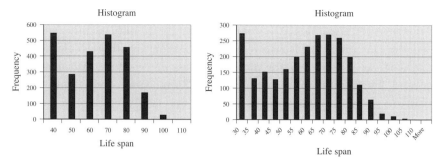

FIGURE 3. Life span in years of the male English gentry. The same data from Guy (1846) are displayed with different bin widths in the form of a histogram. In the histogram on the left, the bin width is 10 years and in the histogram on the right, the bin width is 5 years.

and $20, 25, \ldots 105$ for the right. The program will automatically create a frequency table. It will create a histogram if the box CHART OUTPUT is checked. The frequencies displayed are the number of observations greater than the left bin boundary and less than or equal to the right bin boundary.

DOT PLOT: For quantitative data, especially if the number of observations is moderate to small ($n \leq 20$), a better graphical display method than a histogram is a dot plot. Whereas in a histogram the observations are categorized (placed in bins) and thus summarized, a dot plot displays each data point separately and conveys more information to the reader. In a dot plot, each data point is represented by a dot which is placed either vertically or horizontally on an axis representing the measurement values (Fig. 4).

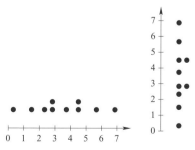

FIGURE 4. Two dot plots of the same ten observations plotted horizontally (left) and vertically (right). Repeated observations of the same value are plotted above or beside each other.

SCATTER PLOT: The most convenient graphical display for two quantitative variables is a scatter plot. In it, one variable is plotted on the horizontal x-axis and the other variable is plotted on the vertical y-axis. Scatter plots allow visual overview of center and spread for each variable while at the same time giving important clues to possible relationships between the two variables.

> **In Excel:** Write your observations into two columns of an Excel spreadsheet. Observations on the same subject should appear in the same row. Click INSERT → CHART... → XY (SCATTER) and highlight your observation columns.

BOX PLOT: Like histograms, box plots display quantitative data. They make use of computed quantities that describe the data (such as the median and quartiles) and display them graphically. In this way, they summarize the data and are thus suited to describe outcomes of experiments in which the number of observations is too large to report each outcome separately (as is done in a dot plot). Box plots are especially useful if the graph is intended to compare measurements on several populations or compare the reaction of a population to different conditions.

To construct a box plot, compute the minimum, maximum, and median as well as the first and third quartiles for each data set that you want to display. Usually, box plots are drawn with the measurements on a vertical scale, but they could be drawn horizontally as well. Label the vertical axis with the variable that was measured and its units. The box plot has a (thick) line cutting through the box that represents the median. The box spans from the first to the third quartile of the observations and "tails" or "whiskers" extend to the maximum and minimum observations, respectively.

In a MODIFIED BOX PLOT, outliers are indicated as stars and the tails are only extended to the largest (smallest) observations which are not outliers.

Example 3.7

For William A. Guy's data set (Guy 1846) which contains observations on the life span of the English upper class in the 19th century, it may be of interest to compare the life spans of the gentry, aristocracy, and sovereignty. Do sovereigns live longer than aristocrats? To be able to make this comparison in a graph, we draw a side-by-side box plot for the available observations on the three groups.

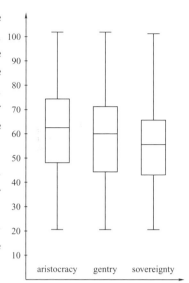

From the graph, it is possible to observe that the life spans do not differ very much between the three groups. Overall, sovereigns seem to live a little less long than the members of the other two groups.

A side-by-side box plot is one example for a graphical display of more than one variable. In Example 3.7, the variables are LIFE SPAN, which is quantitative, and GROUP, which is categorical and takes on values aristocracy, gentry, and sovereignty.

3.3.1   How to Choose the Appropriate Graphical Display for Your Data

The purpose of graphically displaying data is to convey information to the reader. Depending on the aspect of the data on which the author wants to focus, different display methods may be better suited to bring the point across. Generally, it is a good idea to present a more comprehensive overview of the data unless it makes the display overly cluttered. For example, if the intent of the author is to compare repeated measurement values for two groups, a graph that displays not only the means, but also some measurement of the spread, is clearly preferable to one that displays only the means (Fig. 5).

For very large data sets, some form of summarization is necessary. Quantitative data may be displayed in histograms or box plots. Both convey a sense of center and spread of the observations to the reader. If unusual observations (outliers) are important, then a modified box plot may be the best display. Box plots are better suited than histograms to visually compare several populations.

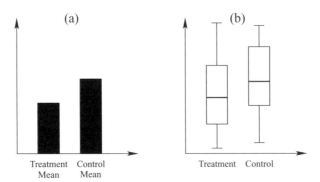

FIGURE 5. Comparing graphical representations of measurements on a treatment and a control population. While it is possible to graph the two population means as bars (a), this display method does not convey enough relevant information to decide whether there is a difference between the two groups. If the same data are displayed in the form of a side-by-side box plot (b), the difference in center can be related to the spread in the observations and a decision of whether or not there is a significant difference can be made.

Quantitative data can be displayed interchangeably with either a bar graph or a pie chart. If quantitative measurements for different populations are to be visually compared, then bar graphs may be plotted side by side with different colors for different populations. This method is reasonable, as long as the number of populations to be compared is small (≤ 3).

3.4   Probability Distributions

Recall that a random variable is a placeholder for the outcome of an experiment which is subject to chance. Before the experiment is conducted, the value of the random variable is unknown. But the probabilities with which the random variable takes on specific values may be known. Statisticians call this the PROBABILITY DISTRIBUTION of the random variable.

Example 3.8

A cross is performed between two individuals who are heterozygous (Aa) for a particular allele. The genotype of the offspring can be of the form AA, Aa, or aa. Before the cross is performed, we do not know which genotype the offspring will have. However, we do know that the probabilities with which these genotypes occur are 0.25, 0.5, and 0.25, respectively.

There are a number of special probability distributions which arise frequently in biological applications. Two of the most important distributions, the BINOMIAL and the NORMAL distributions, are discussed below.

3.4.1   The Binomial Distribution

Consider the following situation: A series of n independent trials is conducted. Each trial will result either in a success (with probability p) or in a failure (with probability $1 - p$). The quantity that we are interested in is the number of trials which will result in a success.

$$X = \text{ number of successes in } n \text{ trials.}$$

This quantity is a random variable, since the outcomes of the trials are subject to chance. It may take on values 0 (in the case that no trial results in a success), 1, 2, etc., up to n (in the case that every trial results in success). A random variable of this kind is called a binomial random variable with parameters n and p. The fundamental assumptions for this kind of random variable are

- The n trials are identical and independent (their outcomes do not influence each other).
- The probability p of a success is the same in each trial.

For every possible value $(0, 1, \ldots, n)$ of a binomial random variable, a probability can be assigned (Fig. 6). For example, the probability of observing zero successes in n trials is $(1-p)^n$, which is the probability

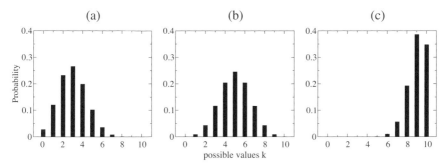

FIGURE 6. Probability histogram for three binomial distributions with parameters $n = 10$ and $p = 0.3$ (a), $p = 0.5$ (b), and $p = 0.9$ (c), respectively. The height of a bar indicates the probability of seeing this many successes in the $n = 10$ trials.

that all trials result in failures. More generally, the probability of seeing k successes in n trials can be computed as

$$P(X = k) = \binom{n}{k} p^k (1 - p)^{n-k}.$$

Here, $\binom{n}{k}$ is a binomial coefficient which may be computed (by hand) as $\binom{n}{k} = n!/k!(n - k)!$ or by using the Excel command "COMBIN(n,k)." Sometimes it is valuable to compute probabilities of the form $P(X \leq 3)$ or "what is the probability of having at most three successes." Statisticians call these expressions CUMULATIVE PROBABILITIES, since probabilities of possible values are accumulated, e.g.,

$$P(X \leq 3) = P(X = 0) + P(X = 1) + P(X = 2) + P(X = 3)$$

In Excel: To compute probabilities or cumulative probabilities for a binomial random variable in Excel, click on any empty cell and type "BINOMDIST(k, n, p, CUMULATIVE)." Here k is the number of successes for which you want to compute the probability, n is the total number of trials conducted, p is the success probability in each trial, and CUMULATIVE takes on the values FALSE or TRUE depending on whether you want to compute the probability of a specific event (e.g., $P(X = 3)$) or a cumulative probability (e.g., $P(X \leq 3)$).

Example 3.9

If two carriers of the gene for albinism have children, then each of their children has a probability of 0.25 of being albino. Whether or not subsequent children are albino is independent of their siblings. The number of children with albinism in this family is thus a binomial random variable with $p = 0.25$ and $n =$ number of children in the family.

To find the probability that in a family with three children in which both parents carry the gene for albinism there is exactly one child with albinism, we use the binomial distribution. Our binomial random variable here is $X =$ number of children with albinism, and we need to compute the probability that X is equal to one.

$$P(X = 1) = \binom{3}{1}(0.25)^1(0.75)^2 = 0.421875$$

or in Excel $P(X = 1) = \text{BINOMDIST}(1, 3, 0.25, \text{FALSE})$
$$= 0.421875.$$

To find the probability that in the same family there are at most two children with albinism, we use the cumulative distribution function instead. For the same random variable X as above, we now want to compute the probability that X is at most 2 (or $X \leq 2$). We can do this by either computing the probabilities that X is equal to 0, 1, or 2 and adding them up, or by using Excel to compute the cumulative probability

$$P(X \leq 2) = \text{BINOMDIST}(2, 3, 0.25, \text{TRUE}) = 0.984375.$$

3.4.2   The Normal Distribution

The distributions of discrete random variables, such as the binomial, may be displayed in histograms (compare Fig. 6). The height of the bar for each possible value corresponds to the probability with which this value will occur in the experiment. For continuous random variables, which may take on any value in an interval, an analogous display method is used (compare Fig. 7).

The most often used continuous distribution in practice is the normal distribution. It arises naturally (but not automatically) in many situations. It also plays an important role whenever statistics are computed based on a large number of observations.

SHAPE: Normal distributions have the characteristic bell shape. They are symmetric and characterized by their mean μ and standard deviation σ, where μ and σ are called the parameters of the normal distribution. All normal distributions have essentially the same shape

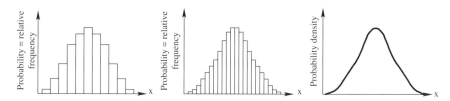

FIGURE 7. Discrete probability distributions can be displayed in histograms. The number of bars in the histogram corresponds to the number of possible values that the random variable may take on (left and middle). For a continuous random variable which can take on infinitely many possible values, the probability distribution is displayed as a density curve (right).

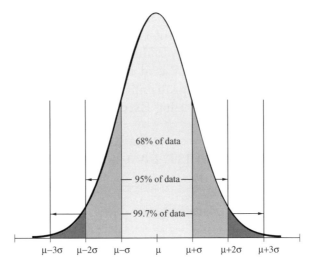

FIGURE 8. The normal distribution is symmetric around its mean μ. Approximately 68% of the data can be found within one standard deviation of the mean ($\mu \pm \sigma$). 95% of the data lie within two standard deviations of the mean ($\mu \pm 2\sigma$) and 99.7% lie within three standard deviations of the mean ($\mu \pm 3\sigma$).

(Fig. 8), which is described by the function

$$f(x) = \frac{1}{\sigma\sqrt{2\pi}} e^{-\frac{1}{2}\left(\frac{x-\mu}{\sigma}\right)^2}.$$

The mean μ determines the center (position on the x-axis). The standard deviation σ determines the spread and with it both the height and width of the curve. Larger standard deviations (more spread) mean flatter, wider curves.

> **Note:** A normal distribution with mean $\mu = 0$ and standard deviation $\sigma = 1$ is referred to as a STANDARD NORMAL DISTRIBUTION.

The normal distribution is a continuous distribution, i.e., random variables that are normally distributed can take on any value in an interval. To compute probabilities for such a random variable, one needs to look at the area under the normal distribution curve. The total area under the curve is equal to one. The area under the curve between two values x_1 and x_2 corresponds to the probability that the random variable takes on values between those numbers $P(x_1 \leq X \leq x_2)$ (compare Fig. 9). In many statistical applications we are interested in "tail probabilities" for distributions. That means

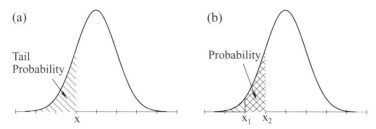

FIGURE 9. (a) Left tail probability for a normal distribution curve. The hatched area corresponds to the probability that the random variable takes on a value less than or equal to x. (b) Probability for a normal random variable. The cross-hatched area corresponds to the probability that the random variable takes on a value between x_1 and x_2. It can be computed by subtracting the left tail corresponding to x_1 from the left tail corresponding to x_2.

we want to find out how large the area under the distribution curve is to the left (right) of a specific value x. Alternatively, we might also need to know which x-value corresponds to a specific (e.g., 0.05) tail probability.

In Excel: To compute probabilities for random variables with a normal distribution, use the Excel command "=NORMDIST(x, μ, σ, CUMULATIVE)." Here μ and σ are the parameters (mean and standard deviation) of the distribution and CUMULATIVE takes on the values "TRUE" and "FALSE" depending on whether you want to compute the area under the curve to the left of x, $P(X \leq x)$ (TRUE) or the height of the curve *at* x, $f(x)$ (FALSE). The second case is very rarely needed in practice.

Example 3.10

To compute the probability that a normal random variable with mean $\mu = 10$ and standard deviation $\sigma = 2$ takes on values between 9 and 13, click on any empty cell in an Excel spreadsheet and type

$$= \text{NORMDIST}(13, 10, 2, \text{TRUE}) - \text{NORMDIST}(9, 10, 2, \text{TRUE})$$

(the answer is 0.62465526). We compute two probabilities and subtract them from each other because we want the probability of values that are to the left of 13 but *not* to the left of 9.

Example 3.11

A normal distribution percentile is the value on the x-axis, such that the tail area under the normal curve to the left of it contains a specific area. For example, to find the 5th percentile of a normal distribution with mean $\mu = 5$ and standard deviation $\sigma = 3$, click on any empty cell in an Excel spreadsheet and type

$$= \text{NORMINV}(0.05, 5, 3)$$

(the answer is 0.0654).

3.4.3   Assessing Normality in Your Data

Normally distributed data is a fundamental assumption in many statistical models. How can an experimenter be convinced that the data collected in an experiment indeed have a normal distribution? Several methods exist: A statistical hypothesis test can answer the question "Are my data normal- Yes/No?" at a desired degree of certainty. Easier visual tools are the so called PROBABILITY (PP) or QUANTILE (QQ) plots. Both plots rely on comparing the observed behavior (distribution) of the data to values that one would expect to see if the data were, indeed, normal.

PROBABILITY PLOT: The expected behavior of a normal distribution is compared to the observed behavior of the data in the form of left tail probabilities. The left tail probability for an observation is the percentage of observations that are smaller than it. For example, for the third smallest of twenty observations this value would be 3/20. This kind of display is also sometimes referred to as a RANKIT PLOT.

QUANTILE PLOT: The normal percentiles (or quantiles) are compared to those of the observed data. Recall that the pth percentile of the observations is the value of the observation such that $p\%$ of the observations are smaller than it. We compute the percentiles of a normal distribution with the same mean and standard deviation as the observed data and compare them to the observations.

QQ-plots are better suited to detect deviation from normality in the tails of data while PP-plots are better suited to detect deviation from normality around the center of the distribution. If the sample size n is large ($n > 30$), there is very little difference between the two methods.

QQ-plot computation in Excel: Suppose you have n observations gathered in an experiment and you want to check whether or not your observations follow a normal distribution.

1. Compute the sample mean $\bar{x}$ and standard deviation s of the observed data values.
2. Write your data values into column A of an Excel spreadsheet and order them by size (smallest first).
3. Assign percentiles to each observation: Write the numbers $1, \ldots, n$ into column B. In cell 1 of column C, type "=(B1−1/3)/(n+1/3)" (plugging in the appropriate value of n). Drag the formula down to fill the column. These are the values to use as percentiles for your observed data.
4. Compute normal percentiles for comparison: type "=NORMINV (C1, $\bar{x}, s$)" into the first cell in column D and drag the formula down to fill the column. Here $\bar{x}$ and s are the sample mean and standard deviation you computed in step 1.
5. Create a scatter plot of the normal percentiles in column D against your original data values (column A).

Interpreting the plot: If your data have approximately a normal distribution then the dots in the scatter plot should lie approximately on the 45° line (see Fig. 10). Slight deviations from the line are to be expected due to variation in the observations. If the scatter plot exhibits any kind of systematic shape (e.g., an S-curve), then this is an indication that the data are not normally distributed.

3.4.4   Data Transformations

If data are not normally distributed, but a statistical model is to be used that requires normality, then sometimes the data can be transformed to satisfy the model assumptions. In this process, a function is chosen and applied to *all* observation values. For example, if one original observation was $x = 5$ and the transformation function is chosen to be $f(x) = \ln x$, then the value $\ln 5 = 1.609$ would be used instead. The transformed values are tested again for normality. If the transformed data pass a normality check, a statistical model can be fitted to the transformed data. If the transformed data fail to meet the normality check, a different transformation may be attempted.

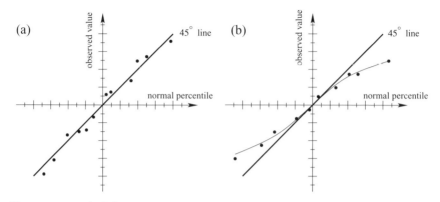

FIGURE 10. A QQ-plot can be used as a visual tool to determine whether data have a normal distribution. If the points in the QQ-plot lie close to the $45°$ line (a), then this is an indication that the data distribution is normal. If the points form any other shape (e.g., an S-curve in b), then this indicates that the data do not have a normal distribution.

Popular functions used for transformations include:

$$\ln x, \quad e^x, \quad \sqrt{x}, \quad x^a$$

Which function is most appropriate for a particular data set is usually determined through a trial and error procedure. Pick a function, transform all data, and check for normality. Repeat as necessary until the transformed data look more or less normal.

Whether or not a transformation is appropriate is a subjective decision. Keep in mind that any transformation makes the statistical model that explains the observed biological process more complex. Simpler transformations are usually preferred over more complicated ones, because they are easier to explain in a biological context. Stay away from convoluted transformations such as $f(x) = \tan^{-1}\left(\frac{2x}{1+e^x}\right)$, even if they produce perfectly normally distributed data. The goal is to achieve a trade-off between normality of the observations and interpretability of the resulting model. Read more on statistical models in Section 4.1.

> **Note:** Sometimes it is *not possible* to find a reasonably simple transformation to make your data look normal. In these cases, the statistical model needs to be adjusted instead of transforming the data. Do not try too hard!

3.5   The Central Limit Theorem

In many biological applications, it is of interest to learn more about a quantitative characteristic of a population. Examples include the average lifetime of a certain organism or the average expression of a particular gene in a plant undergoing a treatment such as drought stress. Other experiments focus on studying the presence or absence of a trait in a population (e.g., What percentage of human patients carry the gene for cystic fibrosis?).

The number of individuals (bacteria, plants, humans) that can be included in a study is usually constrained by the time, money, and manpower available. Even though it is the researcher's intent to draw conclusions about a general POPULATION (such as all humans), one is restricted to work with a SAMPLE (such as the limited number of volunteers in a clinical trial). A different choice of sample will lead to different observations on the subjects involved in the study and thus may lead to different conclusions.

Example 3.12

Suppose that in a population of 10,000 individuals, 1% of subjects carry a particular allele. This particular allele is the focus of study in an experiment. For the experiment, ten individuals are randomly selected from the population to determine whether or not they carry the allele. The number

$$\hat{p} = \frac{\text{number of individuals in the sample who carry the allele}}{\text{number of individuals in the sample}}$$

is called a SAMPLE PROPORTION. Since the allele is relatively rare, recall that only 1% or 100 out of 10,000 individuals carry this marker, most likely none of the ten randomly selected individuals will carry this allele (most likely $\hat{p} = 0$). Of course it is possible that one or even several individuals in the random sample *do* carry the marker (possibly $\hat{p} = 0.1$ or $\hat{p} = 0.2$). The quantity $\hat{p}$ is a random variable (its value depends on the choice of sample). We can compute its distribution using the binomial distribution (compare Section 3.4.1). For example the probability to randomly select ten individuals, none of whom carry the allele ($\hat{p} = 0$), is 0.9043.

> **Note:** When statisticians want to express that a quantity is an estimate
> of a population parameter that was computed based on sample data,
> they often use the "hat" notation. For instance, $\hat{p}$ denotes the estimate of
> the population parameter p, where p for the whole population is usually
> unknown, but $\hat{p}$ can be computed from a set of observations (data).

If we know the true number of individuals in a population who exhibit a characteristic (such as 1% who carry a gene) we can compute
the probability distribution of the random variable $\hat{p}$. However, most
experiments are conducted under circumstances where the population proportion is unknown. Indeed, estimating this population proportion is the very intent of many experiments. Is it still possible to
make statements about the statistical distribution of the sample proportion? Yes! The Central Limit Theorem makes a general statement
about the distribution of the random variable SAMPLE PROPORTION.
Details are discussed in Section 3.5.1.

If an experiment is intended to study a quantitative trait in a population, then an often reported number is the SAMPLE MEAN. That
is, individuals are chosen from the population, the characteristic is
measured on them, and the mean of these measurements is used as
a representative measure for the whole population. Sample means
are also random variables, as their values depend on the choice of
individuals in the sample. In other words, a new sample of the same
size from the same population can produce a different sample mean.
The Central Limit Theorem may also be used to draw conclusions
about the distributions of sample means (see Section 3.5.2).

3.5.1   The Central Limit Theorem for Sample Proportions

Theorem: Suppose that in a large population, p% of individuals exhibit a certain characteristic. A sample of size n is chosen at random
from the population and the sample proportion of the characteristic

$$\hat{p} = \frac{\text{number of individuals exhibiting the characteristic in the sample}}{\text{number of individuals in the sample}}$$

is determined. Then $\hat{p}$ is a random variable whose value depends on
the choice of the sample. If the sample size n is sufficiently large (see
note below), then the distribution of $\hat{p}$ is approximately normal with
mean $\mu_{\hat{p}} = p$ and standard deviation $\sigma_{\hat{p}} = \sqrt{\frac{p(1-p)}{n}}$.

> **Note:** How large should the sample size n be so that the Central Limit Theorem will hold? The answer depends on the frequency of the trait in the population. As a rule of thumb, the sample size n should be large enough so that both $np \geq 10$ and $n(1 - p) \geq 10$. The population from which the sample is chosen at random should be much larger than the sample. This condition is satisfied in most biological applications.

3.5.2   The Central Limit Theorem for Sample Means

Theorem: Suppose that a characteristic has some distribution (not necessarily normal) in a population with mean μ and standard deviation σ. A sample of size n is chosen at random from the population and the characteristic is measured on each individual in the sample: $x_1, \ldots, x_n$. The average of these values $\bar{x} = \frac{x_1 + \cdots + x_n}{n}$ is a random variable whose value depends on the choice of the sample. If the sample size n is sufficiently large (see note below), then the distribution of $\bar{x}$ is approximately normal with mean $\mu_{\bar{x}} = \mu$ and standard deviation $\sigma_{\bar{x}} = \frac{\sigma}{\sqrt{n}}$.

> **Note:** How large should the sample size n be for the Central Limit Theorem to hold? In the case of sample means, the answer depends on the distribution of the characteristic in the population (see Fig. 11). If the distribution is normal, then the Central Limit Theorem holds for sample sizes as small as $n = 2$. The less "normal" the distribution of the characteristic in the population, the larger the sample size n should be for the statement to be valid. In general, sample sizes of $n = 30$ or larger are considered sufficient regardless of the distribution of the characteristic in the population.

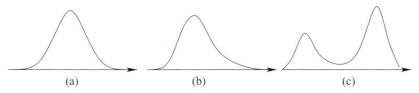

| (a) | (b) | (c) |

FIGURE 11. If a characteristic has a normal distribution in a population (a), then the Central Limit Theorem holds for sample sizes as small as $n = 2$. If the distribution of the characteristic in the population is approximately normal (b), then the Central Limit Theorem holds for small to moderate sample sizes $n \approx 5$. If the distribution of the characteristic in the population is very "non-normal" (c), then the sample size should be large ($n \approx 30$) for the Central Limit Theorem to hold.

Example 3.13

The Central Limit Theorem describes the statistical behavior of a sample mean. To better understand the message behind this important statement, consider the scenario presented in Figure 12. The population here consists of six plants and the characteristic that is measured on them is number of flower petals. The average number of flower petals in the *population* is $\mu = 4$ with a standard deviation of $\sigma = 0.894$. From the population, samples of two flowers are chosen at random $(n = 2)$. Depending on the selection of individuals, the average number of petals $\bar{x}$ in the *sample* varies from 3 to 5. However, the mean of the observed petal averages $\mu_{\bar{x}}$ is equal to 4, the same as the true mean number μ of petals in the population. The standard deviation of the petal means is $\sigma_{\bar{x}} = \frac{\sigma}{\sqrt{n}} = \frac{0.894}{\sqrt{2}} = 0.632$.

In actual biological experiments, the population is likely much larger than the one in Example 3.13, e.g., all *Arabidopsis* plants. The sample size in the real experiment is the number of plants in the laboratory on which the researcher obtains measurements. It may be as small as the one used here $(n = 2)$ or larger. Most likely, the trait studied is more complex than simply the number of flower petals. If the trait can be described numerically, for example, a gene expression value or the physical reaction of a plant to a treatment, then the Central Limit Theorem applies. The researcher uses the measurements on the individuals in the sample to draw conclusions about the general population of *Arabidopsis* plants. Knowing the behavior of a sample statistic allows us to make these generalizations.

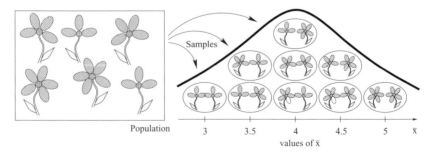

FIGURE 12. Illustration of the Central Limit Theorem.

3.6   Standard Deviation versus Standard Error

Two quantities that are commonly confused are the STANDARD DE-VIATION and the STANDARD ERROR. They model different phenomena and thus are not interchangeable quantities.

Standard Deviation: Describes the variation in observed values within a population. If all individuals in the population exhibit similar values for a certain trait, the standard deviation σ of the trait in the population is small (compare Section 3.2.2).

Standard Error: The standard error $\sigma/\sqrt{n}$ is the standard deviation of the sample statistic $\bar{x}$. Suppose that a sample of size n is taken from the population and the mean of that sample is computed. Suppose further that this process is repeated many times with different samples to yield many sample means of the same quantity. The standard error represents the variation in mean values around the true population mean resulting from choice of sample. As the sample size n increases, the accuracy in representation of the population mean by the sample mean increases and the standard error decreases.

Example 3.14

Suppose the variable of interest is body weight (in g) of mice on different diets. Mice on high-fat diets will typically have higher body weights than mice on low-fat diets. However, due to biological variation, the weights of individual mice will vary (even if the mice are presented with the same kind and amounts of food). For the mice receiving the high-fat diets, the weights will vary more than for the mice on low-fat diets.

To obtain accurate measurements of average weight under both diet conditions, more mice on the high-fat diet should be included in the experiment to make the standard errors for both groups of comparable magnitude. The standard error, which is the discrepancy between average weight and average weight observed in the experimental diet group, decreases with sample size.

3.7   Error Bars

Using the information that the Central Limit Theorem provides about the statistical behavior of a sample statistic such as the mean

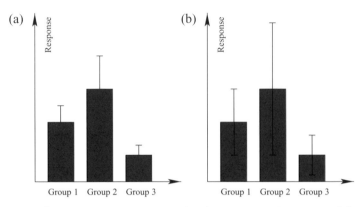

FIGURE 13. In an experiment a quantitative response is measured for three groups. In each group, measurements are taken on four individuals and the observations averaged. The same results can be presented as a bar graph with error bars representing the standard error (a) or as a bar graph with error bars that represent the standard deviation (b). Figure (b) is more informative since it is independent of the sample size and provides a more accurate representation of the variation of the measurements among the observations. The error bars in (b) are twice as wide ($n = 4, 1/\sqrt{n} = 1/2$) as those in a.

or sample proportion allows conclusions to be drawn about the reliability of the information gained from a single sample. How well does the statistic (mean, proportion, etc.) that is computed from the sample data actually represent the population? Error bars represent the variation in the experiment due to sampling and convey to the reader how the reported values may change with a different choice of sample.

Scientists have a choice of ways to convey this information to their audience. The two most popular versions, which are depicted in Figure 13 are as follows:

- Standard Deviation: Let the error bars represent the estimated variation *in the population*. Even though this quantity is estimated from a particular observed sample, it is applicable to the general case of samples of any size. The lengths of the error bars are equal to the standard deviations s of the observations.

- Standard Error: Let the error bars represent the estimated variation *in the sample statistic*. The variation in the sample statistic depends (strongly!) on the size of the sample taken in the experiments. If the experiment were to be repeated with a different number of observations, the error bars would not be directly comparable. The lengths of the error bars are equal to the STANDARD ERROR $s/\sqrt{n}$.

3.8 Correlation

If an experiment records data values for two or more random variables, then one may ask whether there is a relationship between pairs of variables. Does the amount of water that a plant receives influence its height? Or does the amount of light influence height more than the amount of water?

CORRELATION: The statistical measure that describes the relationship between two random variables. Correlation is unitless. It does not depend on the units in which the variables are measured. The correlation coefficient

$$r = \frac{1}{n-1} \sum_{i=1}^{n} \left(\frac{x_i - \bar{x}}{s_x} \right) \left(\frac{y_i - \bar{y}}{s_y} \right)$$

measures the strength and direction of a linear relationship between two variables (Fig. 14). It is also known as Pearson's product moment correlation measure. Here (x_i, y_i) are measurements on n individuals with sample means $\bar{x}$ and $\bar{y}$ and standard deviations s_x and s_y, respectively. The sign of r describes the direction of the association

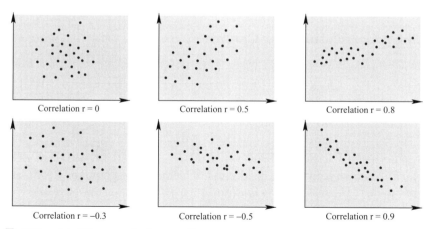

FIGURE 14. The correlation coefficient measures the direction and strength of a linear association between two variables. It is independent of the units that variables are measured in and hence is not related to the slope of the line that interpolates the data points.

between the variables. Positive correlation means that large x-values are associated (on average) with large y-values. Negative correlation means that large x-values are associated (on average) with small y-values. The absolute value of the correlation coefficient r describes the strength of the association. Values close to zero represent weak association while values close to ± 1 represent strong association.

In Excel: To compute Pearson's correlation coefficient for two sets of observations $(x_1, \ldots, x_n)$ and $(y_1, \ldots, y_n)$, write the observations into two columns with values on the same individual in the same row. Click on any empty cell and type "=PEARSON(ARRAY1, ARRAY2)" or "=CORREL(ARRAY1, ARRAY2)" and highlight the respective data arrays.

Note: The correlation coefficient is only able to measure a *linear* relationship between two variables. The variables may have some other relationship (for example, a perfect quadratic relationship, so that for every measured pair of data points it is $y = -x^2$). However, the value of the correlation coefficient of X and Y would not reflect this relationship, as it is not a linear one.

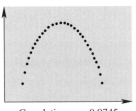

Correlation r = –0.0745

3.8.1   Correlation and Causation

Often the goal of an experiment is to establish proof for some kind of causation. Does second-hand smoke cause lung cancer? Will a particular fertilizer raise the yield in corn? Experimental data are collected (e.g., observations on people exposed to various amounts of second-hand smoke and whether or not they develop lung cancer, or amounts of fertilizer used on a crop and yield of corn). A correlation between hypothesized cause (second-hand smoke, fertilizer) and effect (cancer, yield) is computed. If the association between the variables is high, then causation is often the conclusion. Statistically, however, this conclusion may not be entirely valid, as there can be several reasons for a high association between variables, only one of which is causation.

Suppose two variables (call them X and Y) are studied in an experiment. On a third variable Z, no data are collected. After measuring X and Y on several subjects, the experimenters find an association (high positive or negative correlation) between X and Y. This means that subjects who are found to have high X-levels also tend to have high (or low) Y-levels. How can that be explained?

CAUSATION: One possible explanation for a high correlation between two variables is causation. Elevated levels of variable X cause elevated (or suppressed) levels of variable Y in the subjects. Conversely, high levels of Y may cause high (low) levels of X. The direction of cause and effect cannot be established through statistical correlation.

COMMON RESPONSE: In some cases, an association between two variables X and Y can be observed, because the variables X and Y are both influenced by the third (possibly unobserved) variable Z. In this case statisticians call Z a "lurking" variable.

CONFOUNDING: If the observed variable X is associated with the unobserved variable Z, and X and Z both influence Y, then it is impossible to tell how much of the observed correlation between X and Y is caused by X and how much is caused by Z. In this case, Z would be called a confounding variable.

Example 3.15

It is known that there is a relationship between mood (X) and health (Y) in human patients (Valiant 1998). Optimistic people tend to be in better health than depressed people. It is unclear whether a good mood directly leads to better health (e.g., X causes Y) or whether maybe ill health tends to put people in a bad mood (e.g., Y causes X).

It has been speculated that other factors such as marital stress (DeLongis et al. 1988) or exercise habits (Stewart et al. 2003) may be influencing both mood and health. For example, increased exercise has long been known to lead to better health, but it may also simultaneously improve a person's mood.

Both exercise habits and mood may have an influence on a patient's health. Unless data are collected on the exercise habits as well as the mood of patients in the study, it will not be possible to distinguish between these confounding effects. But even if data are collected on both effects this does not rule out the existence of further confounding variables, such as marital stress.

Example 3.16

Comic Relief: Researchers in Germany have found a high correlation between the number of storks nesting in Brandenburg (the area surrounding the city of Berlin) and birth rates in the same area (Sies 1998; Höfer et al. 2004). Between 1965 and 1980, stork nests have become rarer and, simultaneously, birth rates have become lower. Is this proof for the "Theory of the Stork," as T. Höfer (humorously) asks?

The answer is "no." Even though there may be a high statistical correlation between birth rate and the number of stork nests in this area, this is not proof that babies are delivered by storks. (The authors prefer the more conventional biological explanation.)

An alternative explanation for the high observed correlation lies in environmental factors. As towns become more urban, wildlife moves to less developed areas with more greenery and so do families with young children. Statistically, this would be a "common response" situation. Both stork nests and birth rates are influenced by environmental factors, such as urbanization of formerly rural areas.

Note: It is a mistaken belief that the choice of statistical method, and not the way in which the data are collected, determines whether or not one can conclude causation from correlation. This is not true. Only a carefully planned experiment, which rules out lurking variables or confounding effects is able to establish causation. Thus, establishing high correlation between two variables is necessary but not sufficient to conclude that one condition causes another.

4 Design of Experiments

4.1 Mathematical and Statistical Models

A MATHEMATICAL model is an equation that describes the relationship between two or more variables. Mathematical models are DETERMINISTIC in the sense that if the values of all but one of the variables in the model are known, then the value of the remaining variable can be deduced.

$$\text{Mathematical Model Example: } y = x^2$$

A STATISTICAL model, on the other hand, is not deterministic. If the values of some of the variables in the model are known, then the values of the other variables cannot be computed exactly, although they can be estimated using information from a sample (Fig. 15).

PREDICTOR VARIABLE: If the value of a variable can be determined by the experimenter, this variable is referred to as a predictor variable (e.g., organism type, light level, temperature, or watering in plant growth experiments). Predictor variables are also sometimes referred to as independent or explanatory variables.

RESPONSE VARIABLE: A variable measured in the experiment whose reaction to different levels of predictor variables is to be studied is

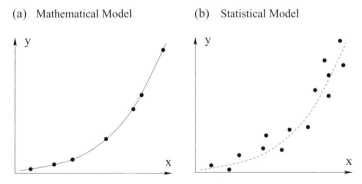

(a) Mathematical Model   (b) Statistical Model

FIGURE 15. The observations are pairs (x, y) where the predictor is denoted x and the response variable is denoted y. The data are plotted as dots for a mathematical (a) and statistical (b) model.

called a response variable (e.g., height or vigor of plants in growth experiment under different conditions). Response variables are also sometimes referred to as the dependent variables.

Example 4.1

Most experiments collect information on several predictor variables and one or more response variables. In a plant growth experiment, temperature, soil moisture, and light conditions could be predictor variables, and plant height or biomass could be measured as response variables representing vigor.

A statistical model describes the distribution of the response (y) for a particular value of the predictor (x). The observed and recorded measurements are understood as observations on a random variable whose distribution may change according to the external conditions.

Statistical Model Example: $y = x^2 + \epsilon$, where $\epsilon \sim \text{Normal}(0,1)$

Here, the response y is represented as a normal random variable with mean x^2, which depends on the predictor variable and variance $\sigma^2 = 1$.

Note: The notation $\epsilon \sim \text{Normal}(0,1)$ stands for "the random variable ϵ has a normal distribution with parameters 0 and 1." In general, statisticians use the symbol $\sim$ to express the distribution of a variable. The distribution is identified by name (here normal) and the values of all parameters of the distribution are identified.

4.1.1 Biological Models

The goal of many biological experiments is to develop an abstract model that explains a behavior or reaction to external influences. The responses can be as simple as the growth of plants under different light conditions or as complex as a cell cycle.

Even in very simple experiments, there are many other factors beyond the predictor and response variables that provide information. Data are not necessarily collected on all of them. For example in the plant growth experiment, every plant is a different individual that may react differently to the same stimulus. It is impossible to keep the experimental conditions (placement on the lab shelf, soil, etc.) exactly the same. And even if the experimental conditions can be very well controlled, other variation introduced by measurement errors or biological differences between individuals still exists. Thus, it is not appropriate to use a deterministic mathematical model to describe the relationship between predictor and response variables. Instead, a statistical model is employed to describe the relationship as precisely as possible, while at the same time allowing for random fluctuations that may be caused by factors not directly observed in the experiment.

4.2   Describing Relationships between Variables

The purpose of a statistical model is to understand the fundamental relationship between variables. This relationship is fundamental in the sense that it is usually not of interest to study how a particular animal (the mouse in your lab cage) or a plant (in your laboratory) behaves, but rather to draw conclusions that are generally valid for a particular strain of mice or a genotype of plant. Since it is not practical or financially feasible to collect data from all mice or all plants, a sample is randomly selected and studied. Data from this sample are then used to choose a statistical model which, under certain conditions, is generally valid for the extended population.

POPULATION: An organism (e.g., mouse) or a group of individuals (e.g., cancer patients) about which you want to draw conclusions.

SAMPLE: A subset of the population from which data are collected.

In a STATISTICAL MODEL, the response variable is represented as a function of the predictor variable plus an error term ϵ whose distribution is known.

$$Y = f(X) + \epsilon$$

If the nature of the functional relationship is known, then the observations are used to "fine-tune" the model. If the functional relationship is unknown, then the observations or data are used to find a functional relationship that provides a good fit.

FITTING a statistical model means selecting a function that relates the predictors to the response. This function should achieve a

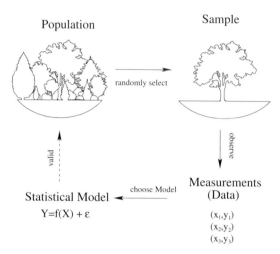

Population   Sample

randomly select

valid   observe

Statistical Model ←—— choose Model —— Measurements
                                        (Data)
Y=f(X) + ε (x_1, y_1)
                                        (x_2, y_2)
                                        (x_3, y_3)

FIGURE 16. A statistical model is fit based on data obtained from a sample. If the sample is representative of the population, then the model is assumed to be valid in general.

trade-off between being as simple as possible and biologically meaningful in the context of the experiment while at the same time explaining as much of the observed variation as possible in the response through the predictors (compare Fig. 16).

Note: In general, statisticians denote unknown random quantities by uppercase letters and observations in experiments or measurements that have already been taken by lowercase letters. The statistical model represents the general population (not just your measurements) and is thus usually formulated in uppercase letters.

Suppose it is known that the predictor variable X and the response variable Y have a linear relationship such that if X increases by a unit quantity, Y will increase on average by a fixed (not necessarily whole unit) quantity.

$$Y = \beta_0 + \beta_1 X + \epsilon$$

In this case, the data will be used to find the most appropriate values for the model parameters β_0 (intercept) and β_1 (slope).

MODEL PARAMETERS: Generally, the function $f(X)$ that is used to model the response Y will depend on one or more PARAMETERS. In the above example of a linear function, the model parameters are β_0 and β_1. These parameters, which describe the general relationship between predictor and response for the entire population (not just the sample), are numbers that are generally unknown. However, their values may be estimated from the sample.

The estimate of a model parameter is called a STATISTIC. Since the value of the statistic depends on the selection of the sample, it is itself a random variable. Usually, model parameters are denoted by Greek letters (α, β, or σ) and their estimates, obtained from a sample of data, are denoted by the corresponding Arabic letters (a, b, or s).

If the functional relationship between a predictor and response is not known, then observations may be used to decide which of several possible candidate models provides a better fit, e.g.,

$$\text{Model 1: } Y = \beta_0 + \beta_1 X + \epsilon, \qquad \text{Model 2: } Y = \beta_0 + \beta_1 X^2 + \epsilon$$

Once an appropriate model has been chosen, the same data may be used to obtain estimates for the parameters in this model.

4.3   Choosing a Sample

What are the conditions under which the observations in a sample lead to a model which can be generalized to the whole population? The general goal is to select a sample which represents the population in as many aspects as possible. Theoretically, this goal is achieved if the sample is taken strictly at random. However, technically this is usually not feasible since there are many factors influencing a population. For example, most often plants to be sampled are grown on the same lot and are not randomly chosen from all possible lots.

RANDOM SAMPLE: Strictly speaking, random sampling means choosing a subset of a population in such a way that every member in the population has the same chance of being chosen for the sample. Individuals must be chosen independently of each other. If strict random sampling is not possible, other sampling strategies exist, but it is important to understand the consequences of employing those strategies.

STRATIFIED SAMPLE: If a population consists of subpopulations which fundamentally differ with respect to one aspect, then individuals can be organized by this aspect into different STRATA. From each stratum, a sample is now selected at random with the intent of representing the overall population as closely as possible in all other aspects.

MATCHED SAMPLING: If an experiment is intended to compare individuals which can be classified into two (or more) distinct groups, then sometimes matched individuals are chosen to differ in the characteristic to be studied but to be as similar as possible in all other aspects.

4.3.1 Problems in Sampling: Bias

A BIASED sample is one which systematically over- or under-represents individuals which exhibit a particular characteristic of interest. Thus, the sample does not accurately reflect the population and any estimation based on such a sample would lead to systematic errors in estimation and erroneous conclusions drawn from the sample.

Example 4.2

Suppose a survey is conducted to determine the prevalence of a particular disease in a human population. Selection bias would occur if the study participants were randomly chosen from hospital patients. If an individual with the disease is more likely than a healthy individual to be treated (for this or some other disease) in a hospital, then hospital patients are more likely to have the disease than healthy individuals. Since we restrict our observations to the hospital group alone, we would thus systematically overestimate the prevalence of the disease.

Selection bias can be corrected if the nature of systematic over- or under-estimation is understood. However, in most applications, this information is not available and requires forethought when designing an experiment.

Example 4.3

Rumor has it that a study is much more likely to be published if the results are statistically significant (i.e., some effect is concluded to exist) than if the results are not statistically significant. This, despite the fact that non-significant results (i.e., studies where we are reasonably sure that there is no effect) may be just as informative for the scientific community, shows that a publication bias exists.

Example 4.4

In microarray experiments, RNA samples are labeled with red (Cy5) and green (Cy3) fluorescent dyes. It is known that the binding affinity of dyes varies with genes. Suppose gene A is more likely to bind with red dye than gene B. Then if a treatment RNA sample is labeled with red and a control RNA sample labeled with green and they are hybridized on an array, the fold-change of gene A (treatment vs. control) can be much higher than for gene B. However, we cannot conclusively say that the treatment is solely responsible for the differential expression of gene A. Because the binding affinity for the two genes differs, the differential expression may also have been caused by the dye bias. To resolve this common problem, many microarray experiments are conducted in the form of a dye swap. A second hybridization to another array is performed in which treatment is now labeled with green and control labeled with red. Only the combined data from both arrays in the dye-swap experiment make it possible to separate the dye (bias) effect from the treatment effect.

4.3.2 Problems in Sampling: Accuracy and Precision

The accuracy and precision of an estimate depend on the method by which the sample is taken. ACCURACY measures bias—unbiased samples are those with the highest accuracy; PRECISION measures variability in the measurements (Fig. 17). If all the information from each observation is combined, how well does this combined information reflect the population? If the variation between observed individuals is large, the estimate will be less precise than if the variation is small.

Note: An accuracy problem can be resolved by adjusting the sampling technique to obtain an unbiased sample or by incorporating bias information (if available) into the statistical model. A precision problem can usually be resolved by increasing the sample size. For more information on how to select an appropriate sample size, refer to Section 4.5.

Many experiments can contain variation, and therefore bias, from several different sources. Not all variation is undesirable; in fact,

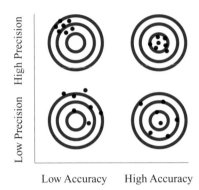

FIGURE 17. Accuracy and precision in measuring a characteristic (bull's-eye) through sample data (black dots).

the variation between two treatment conditions is often the "effect" that the experiment is designed to discover. However, it is extremely important to understand different sources of variation in order to be able to separate the desired variation (effect) from the undesired variation (noise).

BIOLOGICAL VARIATION is introduced through the use of samples from different individuals. Biological variation may be intended (differences between the treatment and control group) or unintended (differences between individuals in the same treatment group). Variation can only be estimated if there is replication. BIOLOGICAL REPLICATION is defined as the repeated analysis using the same method of samples from different individuals (grown, harvested, and measured under the same conditions).

TECHNICAL VARIATION: Even if the same material is analyzed using the sample principle method, the results cannot be expected to be exactly the same due to a variety of possible errors in sample preparation and measurements. Statisticians refer to this as technical variation in the measurements. TECHNICAL REPLICATION may be used to estimate this variation. A technical replicate is defined as the repeated analysis of the same biological sample using the same technique.

Example 4.5

In a microarray experiment, RNA is extracted from two species of *Arabidopsis.* Suppose two plants are selected from each species and the same type of leaf tissue is taken from each plant under the same conditions (A_1, A_2 from species A and B_1, B_2 from species B). In the experiment, the tissue sample from the first plant in each species (A_1, B_1) is color labeled (A_1 red, B_1 green) and hybridized onto one array. If this experiment were repeated with the other two plants (A_2, B_2) using the same color labels and the same type of array, it would constitute biological replication. A dye swap, in which the same biological material (A_1, B_1) is hybridized again onto different arrays with different dye labels, constitutes technical variation.

However, if (A_1, B_1) tissues were labeled red and green, respectively, and (A_2, B_2) tissues were labeled green and red, this would constitute neither biological nor technical variation. Even though there are two observations, we would not be able to separate the variability introduced through different individuals from the variability introduced through the technology (in this case, the dye labeling).

4.4   Choosing a Model

Which statistical model is appropriate for a given set of observations depends largely on the type of data that has been collected. Usually, observations are collected on one or more predictor variables and one or more response variables per experiment (Section 4.1). Each of these variables can be either quantitative or categorical (Section 3.1). There are four possible combinations of types of predictor and response variables:

QUANTITATIVE RESPONSE WITH QUANTITATIVE PREDICTORS

The statistical model for a quantitative response with one or several quantitative predictors is called a REGRESSION MODEL (Section 7.1). A regression model can also be fit if some (but not all) of the predictors are categorical.

CATEGORICAL RESPONSE WITH QUANTITATIVE PREDICTORS

The statistical model for a bivariate categorical response (e.g., Yes/No, dead/alive) and one or more quantitative predictors is called a LOGISTIC REGRESSION MODEL (Section 7.1.3). This model can be extended for situations in which the response variable has more than two possible values (MULTIVARIATE LOGISTIC REGRESSION).

QUANTITATIVE RESPONSE WITH CATEGORICAL PREDICTORS

The statistical model to compare a quantitative response across several populations defined by one or more categorical predictor variables is called an ANOVA MODEL (Section 7.2).

CATEGORICAL RESPONSE WITH CATEGORICAL PREDICTORS

The statistical tool to record observations on categorical predictor and response variables is called a CONTINGENCY TABLE (Section 6.2.5). It can be used to draw conclusions about the relationships between variables.

4.5 Sample Size

According to the Central Limit Theorem (Section 3.5), the standard error for a statistic based on a sample of size n decreases by a factor of $\sqrt{n}$ as the sample size is increased. Larger samples lead to more precise estimates. But how large should a sample be in order to detect some existing effect? The answer depends very much on the particular situation. There are several aspects that influence the choice of an appropriate sample size:

- Effect size
- Significance level
- Variability in population

EFFECT SIZE is the difference between the control and treatment group that you are trying to detect. For example, do colds heal faster if you take a daily vitamin C supplement? In this case the effect size would be the mean difference in duration of a cold with and without the supplement. If this difference is small (6 days without supplement and 5.79 days with supplement), then you will need a large sample to conclude with high statistical significance that there is a difference. If the effect size is large (6 days without supplement and only 3 days with supplement, on average), then a much smaller sample size is sufficient to find the existing effect statistically significant.

SIGNIFICANCE LEVEL: Decisions about populations that are based on samples are always subject to mistakes. To keep the probability of making an erroneous conclusion small, one has to be conservative before declaring an effect significant. Most often the chance of making a wrong decision is bounded by the threshold of 5% (even though

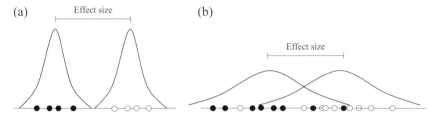

FIGURE 18. If the variability within treatment group (red dots) and control group (white dots) is small (a), it is easier to detect an effect of fixed size than if the variability within a group is large. If the variability is high (b), the sample size needs to be large to distinguish both groups.

other thresholds can and should be used if appropriate). Depending on how certain you want to be that the conclusion drawn from your sample is actually correct, you will have to adjust the sample size. More accuracy in the decision requires larger samples.

THE VARIABILITY of the trait in the population cannot be influenced by the experimenter. If you are measuring a trait whose distribution in a population is very narrow (almost everybody has the same value with very little variation), then only a few observations are needed to draw meaningful conclusions. If the variation of a trait in the population is high, then you will need a larger sample size to draw the equivalent conclusion at the same significance level (Fig. 18).

Calculating exact sample sizes depends on the issues mentioned here, as well as on the data distribution and the scientific question the experiment is designed to answer. Some examples of sample size calculations can be found in Chapters 5 and 6.

4.6   Resampling and Replication

We have already discussed the statistical merits of replication (Section 4.3). Replication means collecting multiple observations under the same experimental circumstances. In a study that includes replication, it is possible to identify sources of variation, estimate the magnitude of variation, and include the variation in the statistical model.

Replication allows the experimenter to understand and describe the variation in the population being studied. The statistical method of RESAMPLING is fundamentally different from replication. Resampling allows one to better understand the behavior of a statistic that is calculated based on a sample. In resampling, data points are randomly selected from the available data set of size n and a model is fit based on the selected subset. Observations may be selected with replacement; this means that the same observation may potentially be selected more than once. This process is repeated many times, and each time the selected data subset may differ, which will lead to different estimates for all of the model parameters. Resampling makes it possible to study the behavior (statistical distribution) of the model parameters based on only one data set.

Example 4.6

A study has been conducted to determine the average width of hummingbird eggs. Five eggs are selected at random and their widths (at the widest part of the egg) are measured (in mm):

$$9.5 \qquad 10.2 \qquad 10.1 \qquad 9.8 \qquad 9.6$$

Estimating the average width of the eggs in the population based on this single sample ($\bar{x} = 9.84$ mm) provides some information.

However, this one number does not tell us how the statistic (i.e., the "sample average") behaves as a random variable. To learn about the distribution of a statistic, we can resample the data (with replacement) many times to create multiple data sets of the same size ($n = 5$). To study the behavior of the sample mean, we compute the estimate "sample mean" based on each resampled data set and then look at the distribution of those estimates. For instance, take 1000 resamples of size five from the original five observations on hummingbird egg width:

Resample	Selected values					Mean
1	9.5	9.8	10.2	9.8	9.5	9.76
2	10.1	10.1	9.8	10.2	10.2	10.08
3	10.2	9.6	9.5	10.2	9.8	9.86
$\vdots$			$\vdots$			$\vdots$
1000	9.8	9.6	9.6	10.1	10.1	9.84

The mean of each resample is computed. A histogram of the resulting 1000 sample means (Fig. 19) shows the distribution of the statistic "sample mean" for these data. As a result, not only can the average

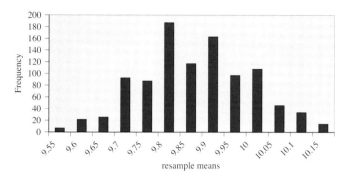

FIGURE 19. Histogram of the sample means computed for 1000 randomly resampled subsets of data from the five original observations.

egg width be estimated, but it is also possible to attach a level of certainty to this information.

Resampling with replacement from the original data is the simplest version of a statistical method called the BOOTSTRAP. The purpose of the Bootstrap method is to derive properties of an estimate (such as a sample mean) from a single sample when it is not possible to take repeated samples from the population.

5 Confidence Intervals

Statistical inference means drawing conclusions from data. Confidence intervals are one important aspect of statistical inference. In general, there is usually a population that we want to study (e.g., humans, *Drosophila*, *Arabidopsis*) and a sample selected from the population. Using the sample, we want to learn more about a population parameter (e.g., the average reaction of humans to a drug, the average wing span of *Drosophila* flies, the percentage of *Arabidopsis* plants which exhibit a genetic marker).

In an experiment, measurements are taken on the sample (drug reaction, wing span, genetic marker). These measurements are usually summarized in the form of a statistic that estimates the population parameter as well as a confidence interval which expresses the certainty of the estimate.

Recall: Population parameters are in general denoted by Greek letters (μ for mean, σ for standard deviation) and the corresponding statistics (calculated from the sample data) by Arabic letters $\bar{x}$ for mean and s for standard deviation.

Due to different sample choices, the value of the statistic calculated from the data may vary. The goal of statistical inference is to draw conclusions about the population parameter from statistics calculated from the data. Most statistics (e.g., sample mean $\bar{x}$, sample standard deviation s, etc.) are chosen because their values computed from data are expected to be close to a true but unknown population parameter (e.g., mean μ, standard deviation σ). But how close (Fig. 20)? Statistics can help to answer this question. Understanding the behavior of sample statistics allows us to quantify how far from the true population parameter we can expect a statistic to be.

Example 5.1

Suppose that we make n random observations on a variable which we know has a normal distribution with unknown mean μ but known standard deviation $\sigma = 2$. In real biological applications, population

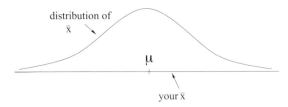

FIGURE 20. In most samples, the calculated value of the sample average $\bar{x}$ will be close (but not equal) to the true unknown population mean μ. The behavior of the statistic $\bar{x}$ is described by the Central Limit Theorem (Section 3.5).

distributions and population standard deviations are very rarely known in practice. Other methods exist to compute confidence intervals in those cases (Section 5.2). However, this simplified example demonstrates the underlying idea behind a confidence interval.

According to the Central Limit Theorem, the average $\bar{x}$ of the n observations has a Normal distribution with mean μ and standard deviation $\sigma/\sqrt{n}$.

$$\bar{x} \sim \text{Normal}(\mu, \sigma^2/n)$$

This knowledge allows us to compute the typical distance of $\bar{x}$ from μ. Of course, the distance of $\bar{x}$ from μ is the same as the distance of μ from $\bar{x}$. Very rarely will the sample average be very far away from the true mean μ. In fact, we can use this property of the sample mean to conclude that the standardized distance of $\bar{x}$ and μ has a standard normal distribution

$$Z = \frac{\bar{x} - \mu}{\sigma/\sqrt{n}} \sim \text{Normal}(0, 1)$$

and we can use properties of the standard normal distribution to determine that the area under the standard normal curve between the values -1.96 and 1.96 has area 0.95.

$$P(-1.96 \leq Z \leq 1.96) = 0.95$$

Combining the above two statements

$$P\left(-1.96 \leq \frac{\bar{x} - \mu}{\sigma/\sqrt{n}} \leq 1.96\right) = 0.95$$

leads to a 95% confidence interval for the population mean μ.

$$\mathrm{CI}_{\mu} = \left[\bar{x} - 1.96\frac{\sigma}{\sqrt{n}}, \bar{x} + 1.96\frac{\sigma}{\sqrt{n}}\right]$$

In practice, when the population distribution and/or the population standard deviation σ are unknown, this confidence interval can be modified to allow for estimation of σ.

To be able to draw general conclusions about population parameters, it is necessary to make assumptions about the data. In the example above, it was assumed that the data are normally distributed and that the standard deviation σ is known. Other common assumptions are random choice of sample and large sample size. Different assumptions on the population from which data are drawn will lead to different formulations of confidence intervals.

5.1 Interpretation of Confidence Intervals

The purpose of a confidence interval is to convey not only an estimate of a population parameter, but also the quality of the estimate. In this sense, confidence intervals are very similar to error bars that are typically used in graphical displays of quantitative data (Section 3.7). However, the interpretation of confidence intervals is a little bit tricky. The word *confidence* is not to be confused with *probability*. The confidence level (typically 95%) is not a probability, and rather, should be understood with respect to repeated sampling. Based on any particular sample, the confidence interval may or may not contain the true population parameter. If samples were to be taken repeatedly and 95% confidence intervals computed based on every sample (Fig. 21), then on average, 95% of those confidence intervals would contain the population parameter.

Note: Assume that one sample of data is collected and a 95% confidence interval for the mean is computed based on $\bar{x}$ for this sample. It is tempting (but wrong) to conclude that this confidence interval contains the true population mean μ with probability 0.95. Since the true mean is a (unknown but) fixed number, it will be contained in any interval with probability either 0 or 1. The confidence level (here 95%) refers to the process of repeated sampling.

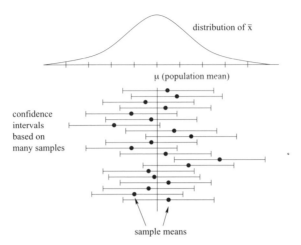

FIGURE 21. Confidence intervals convey the estimate as well as the uncertainty when estimating a particular population parameter (here, the mean μ). The confidence level describes the proportion of samples for which the interval contains the true parameter.

Example 5.2

Many contraceptive methods are advertised by their "effectiveness." What exactly does it mean for a condom, for example, to be 98% effective? It does not mean, of course, that when a condom is used a woman is 2% pregnant after intercourse. As your mother probably told you, you cannot be a little bit pregnant! You either are pregnant, or you are not. In fact, when reading the fine print on a particular condom manufacturer's website, you will find the definition of effectiveness: Out of 100 couples, who all have intercourse on average 83 times a year and use condoms correctly each time they have intercourse, on average only two will become pregnant over the course of the year.

What does this have to do with confidence intervals? As for a confidence interval, the effectiveness rate of 98% (confidence level) does not refer to an individual person (sample). Every time a woman has intercourse and her partner correctly uses a condom, her chances of pregnancy are either 0 (not pregnant) or 1 (pregnant). A confidence interval based on a single sample either does or does not contain the true population parameter. The effectiveness rate or confidence level can only be interpreted as a probability over the span of many

couples or samples. However, in most practical applications, only one sample is collected and observed. Thus, if you are a woman who had intercourse 83 times last year with correct use of a condom every time (and you do not pay particular attention to drastic changes in your body over the course of the year), then at the end of the year you can be 98% confident that you are not pregnant.

5.1.1   Confidence Levels

Most confidence intervals are computed at a confidence level of 95%. This means that if samples are taken repeatedly, then 95% of the confidence intervals based on those samples contain the population parameter. However, there is nothing special about the value 0.95, and confidence intervals may theoretically be computed at any confidence level between 0 and 1. In practice, confidence levels are typically chosen to be between 80% and 99%. Generally, confidence levels are denoted as $1 - \alpha$. For example, $\alpha = 0.05$ for a 95% confidence level. This notation establishes a direct connection to the significance level α in hypothesis testing (Chapter 6).

In general, a higher confidence level makes the confidence interval wider and a lower confidence level makes a confidence interval more narrow. The width of a confidence interval can always be expressed as

$$2 \cdot (\text{critical value}) \cdot (\text{standard error})$$

FIGURE 22. A critical value depends on the distribution of the statistic and the confidence level of the confidence interval. It is the number on the x-axis such that the area under the distribution curve between the positive and negative critical values is equal to the confidence level.

Here, the standard error is the standard deviation of the statistic used to estimate the population parameter (e.g., $\sigma/\sqrt{n}$ for $\bar{x}$) and the critical value is a number that depends both on the distribution of the statistic and the confidence level.

CRITICAL VALUE: In fact, the critical value is the value on the x-axis such that the area under the distribution between the positive and negative critical value is equal to the desired confidence level (Fig. 22).

In Excel: Critical values depend on the distribution of the statistic and the confidence level $1 - \alpha$. For normal distributions, use

$$z_{\alpha/2} = \text{NORMSINV}(1 - \alpha/2)$$

to compute the critical value. For t-distributions, use

$$t_{\alpha/2, n-1} = \text{TINV}(\alpha, n-1),$$

where n is the number of observations from which the confidence interval is computed.

The most commonly used confidence levels are 90%, 95%, and 99%. Critical values for normal and t-distributions at selected sample sizes can be found in Table 1. All other critical values can be computed in Excel using the commands given in the box above.

5.1.2 Precision

Which is better? A 95% confidence interval or a 99% confidence interval? The answer depends on how much precision is required.

TABLE 1. Commonly used critical values for normal and t-distributions at selected sample sizes n.

Distribution	Confidence level		
	90%	95%	99%
Normal	1.64	1.96	2.58
t-distribution, $n = 5$	2.13	2.78	4.60
t-distribution, $n = 10$	1.83	2.26	3.25
t-distribution, $n = 20$	1.73	2.09	2.86
t-distribution, $n = 50$	1.68	2.01	2.68

Higher confidence levels $(1 - \alpha)$ result in wider confidence intervals (less precision).

PRECISION corresponds to the width of the confidence interval. Very narrow confidence intervals at high levels of confidence are the ultimate goal.

There are a number of factors that influence the width of a confidence interval. Their values are usually chosen to achieve a trade-off between confidence (high confidence is preferred) and precision (high precision is preferred).

- The confidence level: Making $1 - \alpha$ larger means making the confidence interval wider since both $z_{\alpha/2}$ and $t_{\alpha/2,n-1}$ increase as α decreases.

- The sample size: A larger sample size n decreases the standard error and thus makes the confidence interval narrower. Notice that increasing the sample size is the only way to provide more precision while keeping the confidence level fixed.

- The variability of the observed variable σ cannot usually be influenced by the experimenter. Nevertheless, it has an influence on the width of confidence intervals. Measurements that are highly variable result in a wider (less precise) confidence interval.

5.2   Computing Confidence Intervals

Theoretically, confidence intervals can be computed for any type of population parameter as long as there is a statistic with which to estimate the parameter and the statistical behavior of this statistic is known. By far the most common cases in practice are confidence intervals for population means or population proportions. For simplicity, we focus here entirely on confidence intervals of these types.

5.2.1   Confidence Intervals for Large Sample Mean

If the sample is taken from the population at random and the sample size is large ($n \geq 30$), then the sample mean has approximately a normal distribution regardless of the population distribution. The population standard deviation σ may be replaced with the sample standard deviation s for very large samples ($n \geq 40$) without changing the distribution of $\bar{x}$.

CONFIDENCE INTERVAL: A $1 - \alpha\%$ confidence interval for the population mean μ is

$$\mathrm{CI}_\mu = \left[\bar{x} - z_{\alpha/2} \frac{s}{\sqrt{n}}, \bar{x} + z_{\alpha/2} \frac{s}{\sqrt{n}} \right]$$

where $\bar{x}$ is the sample mean, s is the sample standard deviation, n is the sample size, and $z_{\alpha/2}$ is the appropriate critical value for the normal distribution (Section 5.1.1).

ASSUMPTIONS: For the above confidence interval formula to be valid, the following conditions must hold.

- The sample is chosen at random from the population (no bias).
- The sample is very large ($n \geq 40$).

Example 5.3

Red yeast rice has been used in China for hundreds of years in cooking and as a food preservative. It contains small amounts of the same substance found in cholesterol-lowering drugs. Researchers conducted a trial to investigate whether the ingestion of red yeast rice (in concentrated capsule form) had a significant effect on the subjects' total LDL cholesterol levels. Fifty-two patients with slightly elevated blood cholesterol levels were chosen for the study. None of the patients took any form of cholesterol-lowering medication (other than the red yeast rice capsules). After a regimen of 15 weeks on the dietary supplement, researchers observed that the average LDL levels in the 52 subjects were reduced by 0.63 points (with a standard deviation of 0.4 points).

This information allows us to compute a 95% confidence interval for the average reduction in LDL cholesterol after a 15-week regimen on the red yeast rice supplement.

$$\text{CI}_\mu = \left[0.63 \pm 1.96 \cdot \frac{0.4}{\sqrt{52}}\right] = [0.521, 0.739]$$

Thus, we can say that we are 95% confident that a 15-week regimen of red yeast rice will reduce the average person's cholesterol level between 0.521 and 0.739 points.

5.2.2 Confidence Interval for Small Sample Mean

Suppose a fairly small sample ($n < 40$) is taken from a normal distribution. The population standard deviation σ is generally unknown and thus needs to be estimated by the sample standard deviation s. In this case, the quantity

$$t = \frac{\bar{x} - \mu}{s/\sqrt{n}} \sim t(df = n - 1)$$

no longer has a normal distribution. The additional uncertainty resulting from the estimation of σ changes the distribution to a Student's t-distribution with $n - 1$ degrees of freedom. t-distributions are shaped very similarly to normal distributions but with slightly thicker tails. Just as the mean and standard deviation determine the exact shape of a normal distribution (compare Section 3.4.2), the number of degrees of freedom (df) determines the exact shape of the t-distribution. For infinite degrees of freedom ($df = \infty$), the t-distribution is identical to the standard normal distribution.

As in the previous case, confidence intervals for the population mean μ can now be constructed that rely only on estimable components. A $(1 - \alpha)$% confidence interval for μ is

$$\text{CI}_\mu = \left[\bar{x} - t_{\alpha/2, n-1}\frac{s}{\sqrt{n}}, \bar{x} + t_{\alpha/2, n-1}\frac{s}{\sqrt{n}}\right]$$

Here, $\bar{x}$ is the sample mean, s is the sample standard deviation, n is the sample size, and $t_{\alpha/2, n-1}$ is the quantile of the t-distribution corresponding to $n - 1$ degrees of freedom (computed as shown in Section 5.1.1).

ASSUMPTIONS:

- The population should be much larger than the sample (rarely a problem in the life sciences).
- The sample must be chosen at random from the population (no bias).
- The observations should be taken independently of each other.
- The measured trait must have (approximately) a normal distribution in the population. This assumption can be verified by one of the procedures to check normality presented in Section 3.4.3.

Example 5.4

Elevated counts of white blood cells in postmenopausal women may be an early warning sign for breast cancer. In a study, white blood cell (WBC) counts were determined for six menopausal women who had recently received a diagnosis of breast cancer but had not yet begun any treatment.

$$\text{WBC counts:} \quad 5 \quad 9 \quad 8 \quad 9 \quad 7 \quad 9$$

A 95% confidence interval for the average white blood cell count in the population of postmenopausal women with breast cancer is

$$\text{CI}_\mu = \left[\bar{x} - t_{\alpha/2,n-1} \frac{s}{\sqrt{n}}, \bar{x} + t_{\alpha/2,n-1} \frac{s}{\sqrt{n}} \right]$$

$$= \left[7.833 - 2.57 \frac{1.602}{\sqrt{6}}, 7.833 + 2.57 \frac{1.602}{\sqrt{6}} \right] = [6.152, 9.514]$$

The upper range of a normal WBC count is around 7. This study finds that the population average for postmenopausal cancer patients lies between 6.152 and 9.514 (at confidence level 95%). This is not enough evidence to conclude that the cancer average is strictly higher than the normal group's average (we have no healthy patients with which to compare in this study). It is also not enough to conclude that the cancer group's average is strictly higher than a 7 WBC count, because the confidence interval includes that value.

To obtain a positive result, the researchers could increase the sample size in their study. This would make $\sqrt{n}$ in the denominator larger and thus the confidence interval narrower. Alternatively, they could also obtain WBC measurements from healthy women as a control and compare confidence intervals for the two groups.

What was actually done in this case is that the experimenters focused solely on very high and very low WBC counts. They obtained counts from both healthy and cancer patients and compared the *proportion* of cancer patients in the groups with very low and very high WBC counts. They found that the proportion of cancer cases in the group with very high WBC counts was significantly higher than in the group with very low WBC counts. Read on in Section 5.2.3 to find how to compute confidence intervals for proportions.

5.2.3 Confidence Intervals for Population Proportion

Consider situations where the response variable in which we are interested is not quantitative but rather categorical. What we are interested in is the *percentage* of individuals in the population who fall into a certain category. (Example: the percentage of cancer patients in the population of postmenopausal women with very high WBC counts).

We cannot measure the whole population but we can take a sample of size n instead and count the number of individuals in the sample who exhibit the trait of interest. Here we denote the (unobservable) population proportion by p and the corresponding (measurable) sample proportion by $\hat{p}$. The sample proportion is computed as

$$\hat{p} = \frac{\text{number of individuals in the sample who exhibit the trait}}{\text{sample size}}.$$

Then, a confidence interval at confidence level $(1 - \alpha)\%$ for the population proportion p is

$$\left[\hat{p} \pm z_{\alpha/2} \sqrt{\frac{\hat{p}\,(1 - \hat{p})}{n}} \right]$$

Here, $z_{\alpha/2}$ is the appropriate critical value of the standard normal distribution (computed as shown in Section 5.1.1).

ASSUMPTIONS: For the above confidence interval formula to be valid, the following assumptions must be satisfied:

- The sample must be drawn randomly and independently from a population that is much larger than the sample.

- The sample size must be reasonably large for the approximation to be valid. As a rule of thumb, a sample can be considered large if it contains at least ten individuals with the trait and at least ten individuals without the trait.

Example 5.5

BRCA1 is a gene that has been linked to breast cancer. Researchers used DNA analysis to search for *BRCA1* mutations in 169 women with family histories of breast cancer. Of the 169 women tested, 27 had *BRCA1* mutations. Let p denote the probability that a woman with a family history of breast cancer will have a *BRCA1* mutation. Find a 95% confidence interval for p.

Our sample estimate is $\hat{p} = \frac{27}{169} = 0.1598$. The standard error for $\hat{p}$ is $\sqrt{\frac{\hat{p}(1-\hat{p})}{n}} = \sqrt{\frac{0.1598(1-0.1598)}{169}} = 0.02819$. The confidence interval for p then becomes

$$[0.1598 \pm 1.96 \cdot 0.02819] = [0.105, 0.215]$$

Thus we are 95% confident that the population proportion p is between 10.5% and 21.5%.

5.3   Sample Size Calculations

How large should a sample be so that a study has sufficient precision for its intended purpose? In general, larger confidence levels lead to wider confidence intervals and increased sample size leads to narrower confidence intervals. Sometimes it is possible to compute how large a sample size should be in a study to achieve a confidence interval of specified length for a specified confidence level.

The length of a confidence interval for a population proportion is $l = 2 \cdot z_{\alpha/2} \sqrt{\frac{\hat{p}(1-\hat{p})}{n}}$. Specifying the confidence level fixes $z_{\alpha/2}$. Specifying the length of the confidence interval makes it possible to solve the above equation for the sample size n

$$n = \left(\frac{2z_{\alpha/2}}{l}\right)^2 \hat{p}(1-\hat{p}).$$

If an estimate for the population proportion is available (before the study is actually conducted), then this number may be used for $\hat{p}$. If this estimate is not available, then the fact that $p(1-p) \leq \frac{1}{4}$ may be used to choose a sample of size at least

$$n = \left(\frac{z_{\alpha/2}}{l}\right)^2.$$

Example 5.6

Suppose we want to conduct a study to estimate the percentage p of left-handed people within the population of the U.S. Assume that we have no preliminary data on left-handedness and that we would like to produce a 95% confidence interval that estimates the true population proportion of left handers within 1 percentage point (i.e., the length of our confidence interval for p should be no more than 0.02). How many subjects should we include in our study?

Here we are looking for the value of n so that $2 \cdot z_{\alpha/2}\sqrt{\frac{\hat{p}(1-\hat{p})}{n}} \leq 0.02$. We do not have any information on $\hat{p}$ from a previous study, so we have to use the approximation

$$n \approx \left(\frac{z_{\alpha/2}}{l}\right)^2 = \left(\frac{1.96}{0.02}\right)^2 = 9,604.$$

Thus we should include 9,604 subjects. If we cannot afford to include this many subjects in our study, we have to be more lenient either in the confidence level (smaller confidence level = fewer subjects) or in our expectations on how narrow we would like the interval estimate to be (wider interval = fewer subjects).

6 Hypothesis Testing

6.1 The Basic Principle

Hypothesis testing is an important example of statistical decision making. The goal is to use data from a sample to investigate a claim about a population parameter. In other words, we want to use the data to answer specific questions. The claim to be investigated is phrased in the form of two opposing statements: the null and alternative hypotheses.

The NULL HYPOTHESIS, usually denoted H_0, is a claim about the population parameter. It must always be formulated as an equality. In the life sciences, the null hypothesis is usually the less exciting outcome (i.e., nothing interesting is happening). Examples include:

- The treatment has no effect, or effect$_{trt} = 0$.
- The gene is not differentially expressed, or $e_{treatment} = e_{control}$.

The ALTERNATIVE HYPOTHESIS, denoted H_a, is the opposite of the null hypothesis. This is usually the statement that the scientist really suspects to be true. Alternative hypotheses may be phrased in a one-sided ($<$ or $>$) or a two-sided ($\neq$) form.

Example 6.1

A biologist wants to study whether mild drought has an effect on the average height of tomato plants compared to plant growth under ideal control conditions. She suspects that plants grown under drought conditions will grow to be shorter than the control plants. In this case, the null hypothesis is

$$H_0 : \mu_{\text{drought}} = \mu_{\text{control}}$$

and the one-sided alternative hypothesis is

$$H_a : \mu_{\text{drought}} < \mu_{\text{control}}$$

where the population average plant height is denoted μ.

Example 6.2

In a microarray experiment, differential expression of a gene can be tested by considering the gene's average expression under control and treatment conditions. The null hypothesis is that the expression is (on average) the same under both conditions

$$H_0 : \mu_{\text{treatment}} = \mu_{\text{control}}$$

Assuming there is no prior knowledge of whether the gene is up-regulated or down-regulated, the alternative hypothesis is two-sided and represented as an inequality

$$H_a : \mu_{\text{treatment}} \neq \mu_{\text{control}}$$

Of course, one cannot expect that two gene expression intensities measured with microarray technology will be exactly equal. The purpose of the hypothesis testing procedure is to decide whether the variation observed in the data is consistent with the null hypothesis (i.e., randomness) or whether the information in the data favors the alternative hypothesis. To make this decision, statisticians utilize what they call a test statistic function.

TEST STATISTIC: Any function whose value can be computed from the sample data, and whose theoretical behavior (distribution) is known when the null hypothesis is true, can be used as a test statistic. A collection of common test statistic functions can be found in Section 6.2. Knowing the distribution, or behavior, of a test statistic function allows us to determine which values are likely or unlikely in a random situation. If we observe a test statistic value that is extremely unlikely to occur in a random situation (i.e., if the null hypothesis is true), then we have observed evidence *against* the null hypothesis. In this case, statisticians say they *reject* the null hypothesis in favor of the alternative.

> **Note:** One can reject the null hypothesis (claim of randomness) in favor of the alternative (pattern in data). However, one can never "accept" randomness, since a pattern that looks random may also have resulted from simply not searching hard enough (not collecting enough data). Instead, we say that one "fails to reject" the null hypothesis when there is insufficient evidence for the alternative.

There are many test statistic functions, each developed for a particular set of experimental conditions and to answer a specific question.

Depending on whether the null hypothesis is based on means (or proportions) of two or more groups, a different function will be appropriate. Although you can read more about specific testing procedures in Section 6.2, it is important to realize that every hypothesis test can be accomplished using five common steps.

THE FIVE STEPS OF A HYPOTHESIS TEST

1. Decide on a significance level α (Section 6.1.4).

2. Formulate the null hypothesis and alternative hypothesis.

3. Choose an appropriate test statistic (Section 6.2).

   (a) Know the distribution of the test statistic under H_0 (common testing procedures will always tell you what distribution to use).

   (b) Compute the value of the test statistic from your data.

4. Compute a p-value for your test and compare with α (Section 6.1.1).

5. Formulate a conclusion sentence that you can report in a publication.

6.1.1 p-values

p-VALUE: A number that is often reported with the conclusion of a hypothesis test is the p-value. A p-value is the probability of observing data that are less compatible with the null hypothesis by chance than data that are observed if the null hypothesis were true.

Figure 23 illustrates distribution curves for a test statistic under a null hypothesis with three possible alternative hypotheses. The p-value is the area under the null hypothesis distribution curve starting at the observed test statistic value and extending in the direction of the alternative hypothesis. If the alternative is one sided, the p-value is the area under the curve in one tail of the distribution. If the alternative is two-sided, the p-value is the area under the curve in both tails of the distribution.

The p-value can be used to decide whether or not to reject a given null hypothesis after viewing the data. If the p-value is small, we conclude that the observed data are unlikely to have been generated by a mechanism that conforms with the null hypothesis (i.e., randomness) and we thus reject the null hypothesis in favor of the alternative

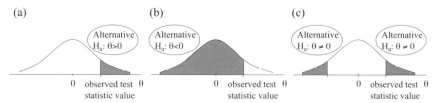

(a)   (b)   (c)

FIGURE 23. Suppose a hypothesis test is designed to test whether some parameter θ is equal to zero (null hypothesis). There are three possible alternatives: (a) $H_a : \theta > 0$, (b) $H_a : \theta < 0$, and (c) $H_a : \theta \neq 0$. The p-value is the probability that a test statistic takes on values *less* compatible with the null hypothesis than the test statistic value observed from the data. This probability (p-value) is the shaded area under the test statistic distribution curve for H_0 in the direction of the alternative hypothesis.

hypothesis. If the p-value is large, we do not have enough evidence *against* the null hypothesis. In this case, statisticians say that they "fail to reject the null hypothesis." How small is small? The p-value is compared against the significance level α (Section 6.1.4) of the test. A p-value is small if $p < \alpha$ and large if $p \geq \alpha$.

6.1.2 Errors in Hypothesis Testing

Two possible mistakes can be made in statistical hypothesis testing. A null hypothesis may be rejected even though it is true. This is known as a Type I error. In practice this kind of error can be interpreted as seeing a significant effect where there is none. A different kind of mistake is made if a null hypothesis is not rejected even though it is false. This is known as a Type II error and corresponds to overlooking an existing effect or missing a significant result.

	Truth	
Test decision	H_0 true	H_0 false
Fail to reject H_0	correct decision	Type II error
Reject H_0	Type I error	correct decision

The significance level α of a test maintains the chance of a Type I error below a level acceptable to the experimenter (Section 6.1.4). For example, if a result is declared statistically significant whenever the p-value is less than $\alpha = 0.05$, then there is a 5% chance of commiting a Type I error every time a test is conducted. Unfortunately, lowering the significance level (to make Type I errors less likely) will increase the probability of committing a Type II error. The only way to lower both Type I and Type II error probabilities is to increase the number of observations in the sample (yes, collect more data).

6.1.3   Power of a Test

The probability of correctly rejecting a false null hypothesis is called the POWER of a test. Practically, power describes a test's ability to correctly identify an existing result. Since this situation is the opposite of committing a Type II error, the power can also be expressed as $1 - \beta$ where β is the probability of a Type II error.

To compute the power of a test, one has to know the distribution of the test statistic under an alternative hypothesis. In some testing scenarios this is easy, but in other situations this can be quite complicated because the distribution of the test statistic under an alternative does not necessarily have the same shape as the distribution of the test statistic under the null hypothesis. If the distribution were known, then the power could be computed by integration as the area under the alternative distribution curve beginning at the observed test statistic value away from the null hypothesis.

6.1.4   Interpreting Statistical Significance

If the p-value of a statistical hypothesis test is smaller than the significance level α of the test, the null hypothesis is rejected in favor of the alternative. Some people find this concept confusing because it does *not* mean that the alternative actually *is* true. Similarly, failing to reject the null hypothesis does not prove that the null hypothesis is true. Statistical hypothesis tests are not proofs in the mathematical sense; they are simply a confirmation or contradiction of some prior belief. The evidence supplied via the data against a null hypothesis needs to be convincing enough (p-value $< \alpha$) before this prior belief is abandoned.

Conventionally, the level of statistical significance is chosen to be $\alpha = 0.05$. However, there is nothing magical about this 5% number. Other meaningful significance levels may be $\alpha = 0.01$ (1%) or even $\alpha = 0.1$ (10%), depending on the application. Thus, whether or not a result is "statistically significant" depends not only on the scientific question and the data, but also on the researcher's willingness to be incorrect based upon a chosen level of statistical significance. For this reason, it is preferable to publish the results of statistical hypothesis tests as p-values rather than as significant or non-significant decisions. p-values contain more information and allow every reader to draw his own conclusions at a meaningful level of significance.

6.2   Common Hypothesis Tests

In this section, we present a selection of commonly used hypothesis testing procedures. In each case, the situation for which the test is designed is described and examples of possible null hypotheses and alternatives are presented. The test statistic is stated together with its distribution. Furthermore, the possible assumptions that need to be satisfied for the testing procedure to be valid are listed.

Review the five steps for hypothesis testing outlined in Section 6.1. Begin by choosing a significance level that you want to work with (Section 6.1.4). Formulate the question you want to ask and select an appropriate null hypothesis and alternative. Depending on the parameter in your null hypothesis (mean or proportion) and the number of populations that you are comparing, identify an appropriate testing procedure (Table 2). The testing procedure that you choose will tell you the test statistic function to use and the distribution of the test statistic function. Compute the value of the test statistic for your data set and obtain the p-value for your test. For many tests, there are Excel commands that skip the test statistic computation step and directly output the p-value for you. Compare the p-value to your chosen significance level and formulate a conclusion in the context of the question being asked.

A final important part of every hypothesis test is to check that your data satisfy the assumptions for the hypothesis testing procedure you chose. Often, these assumptions demand a minimum sample size (which is easy to check). Other times, the assumptions demand normality of the observations which could be checked with a PP-plot or a QQ-plot (Section 3.4.3).

6.2.1 t-test

t-tests are among the most frequently used testing procedures in the biological sciences. They are designed for three distinct applications:

- Comparing the mean of one single population to a fixed constant;

TABLE 2. An overview of common testing procedures described in this step-by step handbook.

	Population parameter of interest			
	Mean	Section	Proportion	Section
One-sample, parameter versus a constant	One-sample t-test	(6.2.1)	One-sample z-test Fisher's exact test	(6.2.2) (6.3.2)
Two independent samples, parameters compared	Two-independent-sample t-test Wilcoxon-Mann-Whitney test Permutation test	(6.2.1) (6.3.1) (6.3.3)	Two-sample z-test Permutation test	(6.2.2) (6.3.3)
Two dependent samples, parameters compared	Paired-sample t-test	(6.2.1)		
Multiple independent samples	F-test Tukey's and Scheffé's tests Permutation test	(6.2.3) (6.2.4) (6.3.3)		
Categorical data	χ^2 goodness-of-fit test χ^2-test for independence Fisher's exact test	(6.2.5) (6.2.5) (6.3.2)		

The name of the testing procedure is listed along with section references for easy access.

- Comparing the means of two independent populations to each other;
- Comparing two dependent measurements (for instance, before and after measurements on the same individuals).

We consider these three different applications separately.

One-sample t-test

The one-sample t-test provides a test of whether the mean μ of a population is equal to a pre-specified constant value δ. Most often, this value is $\delta = 0$, but it does not have to be. Select an appropriate significance level α for your application. The decision will be based on data from a single sample of size n, where $\bar{x}$ denotes the mean of the sample and s denotes the sample standard deviation.

HYPOTHESES:

$$H_0 : \mu = \delta \text{ vs. } H_a : \mu > \delta \text{ (or } \mu < \delta \text{ or } \mu \neq \delta)$$

The null hypothesis is that the population mean μ is equal to δ. The alternative hypothesis states that the population mean is either larger, smaller, or generally not equal to δ. Which alternative hypothesis should be used depends on the application.

TEST STATISTIC:

$$t = \frac{\bar{x} - \delta}{s/\sqrt{n}} \sim t(df = n - 1)$$

Here, the test statistic has a t-distribution with $n - 1$ degrees of freedom.

> **In Excel:** Compute the mean and standard deviation of your sample using the "AVERAGE()" and "STDEV()" commands in Excel (compare Section 3.2.2). Compute the value of the test statistic using the formula above. The square root command in Excel is "SQRT()." Use the command "TDIST(t, df, TAILS)" to compute your p-value. Here, t is the value of the test statistic you computed previously, $df = n - 1$ is the number of degrees of freedom, and TAILS takes on value 1 (for a one-sided alternative) or 2 (for a two-sided alternative).

ASSUMPTIONS:

- The sample must represent the population (no bias).
- The data either need to be approximately normally distributed *or* the sample size needs to be large ($n > 30$).

Two-sample t-test

For two independent populations, the two-sample t-test tests whether the population means μ_1 and μ_2 are equal or not. The decision is made by considering data from samples of size n_1 and n_2 (not necessarily equal). Let $\bar{x}_1$ and $\bar{x}_2$ denote the sample means and s_1 and s_2 the sample standard deviations of the two samples. Choose an appropriate significance level α for your application.

HYPOTHESES:

$$H_0 : \mu_1 = \mu_2 \text{ vs. } H_a : \mu_1 > \mu_2 \text{ (or } \mu_1 < \mu_2 \text{ or } \mu_1 \neq \mu_2)$$

TEST STATISTIC: There are two different test statistics that are used to perform a two-sample t-test depending on whether you assume the two population variances to be equal or not. If the populations are assumed to have equal variance (homoscedastic), use

$$t = \frac{\bar{x}_1 - \bar{x}_2}{\sqrt{\frac{(n_1-1)s_1^2+(n_2-1)s_2^2}{n_1+n_2-2} \cdot \left(\frac{1}{n_1} + \frac{1}{n_2}\right)}} \sim t(df = n_1 + n_2 - 2)$$

If, on the other hand, the two samples cannot be assumed to have been taken from populations with equal variances (heteroscedastic), use the test statistic

$$t = \frac{\bar{x}_1 - \bar{x}_2}{\sqrt{\frac{s_1^2}{n_1} + \frac{s_2^2}{n_s}}} \sim t\left(df = \frac{\left(\frac{s_1^2}{n_1} + \frac{s_2^2}{n_2}\right)^2}{\left(\frac{s_1^2}{n_1}\right)^2 /(n_1 - 1) + \left(\frac{s_2^2}{n_2}\right)^2 /(n_2 - 1)}\right)$$

In Excel: To conduct a t-test for two independent samples in Excel, write your data in two columns, click on any empty cell and type "=TTEST(ARRAY1, ARRAY2, TAILS, TYPE)." Highlight ARRAY1 and then highlight the data column and hit return. Repeat for the second column of data and ARRAY2. TAILS takes on values 1 and 2, depending on whether you are testing against a one-sided ($<$ or $>$) or two-sided ($\neq$) alternative. TYPE takes on values 2 for a two-independent-sample t-test with equal variance assumption and 3 for a two-independent-sample t-test with no equal variance assumption. The Excel output will produce the p-value of the selected test.

ASSUMPTIONS:

- The samples must be independent and represent the unique populations from which they are taken (no bias).

- The data need to be normally distributed in both samples *or* both sample sizes (n_1 and n_2) must be large. Two-sample t-tests are quite robust, which means that sample sizes can be as small as $n_1 \geq 5$ and $n_2 \geq 5$ if the departure from normality is not drastic.

For two populations with non-normally distributed data but with equal variance, there is a non-parametric alternative (i.e., no distribution assumptions are placed on the data) to the homoscedastic two-sample t-test. It is called the Wilcoxon-Mann-Whitney test (Section 6.3.1). This test can be applied to samples of any size, but it is not very powerful for extremely small samples. For moderate-sized samples ($n \geq 10$) with equal variances, a two-independent-sample t-test can also be replaced by a permutation test (Section 6.3.3).

Paired t-test

The two-sample t-test relies on data from two *independent* populations. In many cases, independence is not a meaningful assumption. For instance, measurements may be taken repeatedly on the same individuals under different treatment conditions. To test whether the mean measurements under both conditions are the same, consider the n dependent pairs of observations and compute differences $d_i = x_i^{(1)} - x_i^{(2)}$, where $x_i^{(1)}$ is the observation on individual i under treatment 1 and $x_i^{(2)}$ is the observation on the *same* individual under treatment 2. Let $\bar{d}$ denote the sample mean of the differences and s_d the sample standard deviation of the differences.

HYPOTHESES:

$$H_0 : d = 0 \text{ vs. } H_a : d > 0 \text{ (or } d < 0 \text{ or } d \neq 0)$$

TEST STATISTIC:

$$t = \frac{\bar{d}}{s_d/\sqrt{n}} \sim t(df = n - 1)$$

ASSUMPTIONS:

- Even though the measurements on the same individual are not assumed to be independent, the n individuals should be chosen independently and represent the population they are chosen from.
- Either the two repeated measures should be normally distributed or the number of individuals n must be large ($n > 15$).

In Excel: To conduct a paired t-test in Excel, write your data in two columns, click on any empty cell, and type "=TTEST(ARRAY1, ARRAY2, TAILS, TYPE)." Highlight ARRAY1 and then highlight the data column and hit return. Repeat for the second column of data and ARRAY2. TAILS takes on values 1 and 2, depending on whether you are testing against a one-sided ($<$ or $>$) or two-sided ($\neq$) alternative. TYPE takes on the value 1 for a paired-sample t-test. The Excel output will produce the p-value of the selected test.

Example 6.3

Male *Drosophila* have some of the longest sperm cells of all organisms on earth (including humans). However, the length of the sperm varies by breed. Do males of the species *Drosophila melanogaster* (the so-called fruit fly) and *Drosophila simulans* differ significantly with respect to the length of their sperm cells?

We follow the five-step procedure outlined in Section 6.1. An experiment has produced $n_1 = 15$ observations on *D. melanogaster* sperm. The cells have an average length of $\bar{x}_1 = 1.8$ mm with standard deviation $s_1 = 0.12$ (that is approximately 300 times the length of human sperm cells). There are $n_2 = 16$ observations on *D. simulans* sperm with average length $\bar{x}_2 = 1.16$ mm and standard deviation $s_2 = 0.09$.

1. Select significance level $\alpha = 0.05$.
2. Our null hypothesis is that the average sperm lengths for *D. melanogaster* and *D. simulans* are equal. Assuming that we have no prior information about the flies, we use the alternative hypothesis that the sperm lengths are not equal.

$$H_0 : \mu_1 = \mu_2 \text{ vs. } H_a : \mu_1 \neq \mu_2$$

3. Even though the sample standard deviations are not equal (0.12 and 0.09), it is still reasonable to assume that the population variances are equal (as a rule of thumb, check whether the larger sample standard deviation is no more than twice the smaller). Hence, we will use a two-sample t-test with the equal variance assumption.

4. If you have the complete data set, you could enter it into an Excel spreadsheet and use the command "=TTEST(ARRAY1, ARRAY2, 2, 2)" to perform a two-sample t-test with a two-sided alternative and homoscedastic variance assumption to directly compute the p-value.

   If you do not have the complete data set, but only the summary statistics (from above), compute the test statistic value by hand

   $$t = \frac{1.8 - 1.16}{\sqrt{\frac{14(0.12)^2 + 15(0.09)^2}{29}\left(\frac{1}{15} + \frac{1}{16}\right)}} = 16.87$$

   Compute the p-value for this test using the Excel command

   $$\text{"=TDIST(16.87,29,2)."}$$

   The result is $p = 1.56 \times 10^{-16}$.

5. This p-value is much smaller than α ($= 0.05$); therefore we reject the null hypothesis and instead state that the data contain enough evidence ($p = 1.56 \times 10^{-16}$) to conclude that the average sperm lengths are different.

We do not have to check the data for normality (for example, with a PP-plot) since both samples are reasonably large ($n > 10$).

6.2.2 z-test

Instead of sample means, the z-test compares one or more population proportions, e.g., the percentage of patients who experience a heart attack while on an exercise regimen compared to a control group. To compare one population proportion to a constant or to compare two population proportions to each other, a sample is selected from each population and the sample proportion $\hat{p}$ is calculated as the percentage of individuals in the sample who exhibit the trait. As in the case of the t-test, the z-test exists as a one-sample and a two-sample form.

One-sample z-test

The goal of this procedure is to decide whether or not a population proportion p is equal to a predetermined constant p_0. Suppose a sample of size n is selected from a large population and the sample proportion $\hat{p}$ of individuals which exhibit some trait is observed.

HYPOTHESES:

$$H_0 : p = p_0 \text{ vs. } H_a : p > p_0 \text{ (or } p < p_0 \text{ or } p \neq p_0)$$

TEST STATISTIC:

$$z = \frac{\hat{p} - p_0}{\sqrt{\frac{p_0(1-p_0)}{n}}} \sim \text{Normal}(0, 1)$$

ASSUMPTIONS:

- The population should be much larger than the sample (rarely a problem in the life sciences) and the sample must represent the population (no bias).

- The sample size must be large enough so that $np_0 \geq 10$ and $n(1 - p_0) \geq 10$. Essentially, the sample should contain at least ten observations of each type (with and without the trait).

For smaller samples, an alternative non-parametric test (i.e., no distribution assumptions are placed on the data) exists. It is called Fisher's exact test (Section 6.3.2).

In Excel: Count the number of individuals X that display the trait in your sample. Compute the population proportion $\hat{p} = X/n$ and the value of the z-test statistic using the above formula. The square root command in Excel is "SQRT()." To compute the p-value, use the Excel command "=NORMSDIST(z)" where z is the value of your test statistic. The result will be the area under the normal distribution curve to the *left* of your test statistic. Depending on the formulation of your alternative hypothesis, this may be your p-value or you may have to subtract the result from 1 to yield your p-value (compare Fig. 23).

Example 6.4

A group of fruit flies is conditioned to associate the color red with food. 42 flies were exposed to a maze with two possible options—one path is colored red and the other green. 28 flies chose the red option and the remaining 14 chose green. The hypothesis under investigation asks whether the conditioning worked. Is the number of flies who chose the red option larger than we would expect by chance?

1. Select significance level $\alpha = 0.05$.
2. Let p be the proportion of conditioned flies (not just the flies in our experiment) who choose red when presented with the two options. If the conditioning does not work, we would expect half of the flies ($p = 0.5$) to choose either option. If the conditioning works, then more than half of the flies will choose red.

$$H_0 : p = 0.5 \text{ vs. } H_a : p > 0.5$$

3. The value of the test statistic for this experiment is

$$z = \frac{\frac{28}{42} - 0.5}{\sqrt{\frac{0.5(1-0.5)}{42}}} = 2.16.$$

The test statistic has a standard normal distribution, Normal(0,1).

4. Find the p-value of this one-sided test by computing the right tail area under the normal distribution to the right of $z = 2.16$ using the Excel command "1 - NORMSDIST(2.16)." Recall that Excel computes left tail probabilities; thus the p-value is one minus the left tail area ($p = 0.0154$).

5. This p-value is small ($p < \alpha$); therefore we reject the null hypothesis and conclude that the conditioning worked on the flies.

The assumptions are satisfied, since $(42)(.5) = 21 \geq 10$ and $(42)(1 - .5) \geq 10$. There were at least ten flies who chose either option in the experiment.

Two-sample z-test

Independent samples of size n_1 and n_2, respectively, are drawn from two populations, and the proportion of successes is determined in each sample. The question of interest is whether or not the proportions of successes in both populations are the same.

HYPOTHESES:

$$H_0 : p_1 = p_2 \text{ vs. } H_a : p_1 > p_2 \text{ (or } p_1 < p_2 \text{ or } p_1 \neq p_2)$$

TEST STATISTIC: Let $\hat{p}_1$ and $\hat{p}_2$ denote the sample proportions. Furthermore, let $\hat{p}$ denote the combined relative frequency of successes for both samples.

$$z = \frac{\hat{p}_1 - \hat{p}_2}{\sqrt{\hat{p}(1 - \hat{p}) \left(\frac{1}{n_1} + \frac{1}{n_2} \right)}} \sim \text{Normal}(0, 1), \quad \text{where } \hat{p} = \frac{n_1 p_1 + n_2 p_2}{n_1 + n_2}$$

ASSUMPTIONS:

- Both samples must be chosen independently of each other and should represent their respective populations (no bias).
- The number of successes and failures in both samples should be at least five.

Example 6.5

An experiment is designed to study a candidate gene for alcoholism in mice. 27 field mice and 25 knockout mice for the candidate gene are both offered spiked and virgin drinks. 18 of the field mice and 12 of the knockout mice chose the spiked drink. Is this enough evidence to conclude that the candidate gene is linked to the drink preference of the mice?

1. Choose significance level $\alpha = 0.05$.
2. Let p_1 be the population proportion of field mice (not only those in our lab) that would choose the spiked drink and let p_2 be the population proportion of knockout mice that would choose the spiked drink. The null hypothesis and alternative are
$$H_0 : p_1 = p_2 \text{ vs. } H_a : p_1 > p_2$$
3. From the sampled data, the combined success frequency is
$$\hat{p} = \frac{18 + 12}{27 + 25} = 0.577$$

The value of the z-test statistic is

$$z = \frac{\hat{p}_1 - \hat{p}_2}{\sqrt{\hat{p}(1-\hat{p})\left(\frac{1}{n_1} + \frac{1}{n_2}\right)}} = \frac{\frac{18}{27} - \frac{12}{25}}{\sqrt{0.577(1 - 0.577)\left(\frac{1}{27} + \frac{1}{25}\right)}} = 1.361$$

The test statistic has a standard normal distribution, Normal(0,1).

4. The p-value is the right tail area of the standard normal distribution (in this case, the area to the right of $z = 1.361$). It can be computed using the command "=1 – NORMSDIST(1.361)." Recall that Excel computes left tail probabilities; thus we have to take the "1 minus." The p-value for this example is 0.0867.

5. This value is large (compared to $\alpha = 0.05$). Thus, we fail to reject the null hypothesis and conclude that these data do not provide enough evidence to state that the candidate gene is linked to the drinking preference of mice (i.e., there is no evidence that field mice have a higher population proportion of drinkers than the knockout mice).

The assumptions are satisfied, since there are 18 successes and 9 failures in the field mouse (control) group and 12 successes and 13 failures in the knockout group.

6.2.3 F-test

Recall that the two-sample t-test is used to decide whether the means of two populations are equal or not. In some experimental situations, more than two experimental conditions are considered. The F-test is used to decide whether the means of k populations are *all* equal.

The opposite of this statement is that *at least one* of the populations has a different mean than the others.

HYPOTHESES:

$$H_0 : \mu_1 = \mu_2 = \cdots = \mu_k \text{ vs. } H_a : \text{ at least one } \mu_i \text{ is different}$$

TEST STATISTIC: The test statistic used for the F-test compares the variation within the k groups to the variation among the means of the k groups. Let x_{ij} denote the jth observation from population i $(j = 1, 2, \ldots, n_i)$ and let $\bar{x}_i$ denote the average of the observations from population i. Let $n = n_1 + \cdots + n_k$ denote the total sample size and $\bar{x}$ the (grand) average of all n observations.

$$F = \frac{\frac{1}{k-1} \sum_{i=1}^{k} n_i (\bar{x}_i - \bar{x})^2}{\frac{1}{n-k} \sum_{i=1}^{k} \sum_{j=1}^{n_i} (x_{ij} - \bar{x}_i)^2} \sim F(k-1, n-k)$$

If the variation among the means is large compared to the variation within the groups, the F-test statistic is large. In this case the corresponding p-value is small and the null hypothesis of equal means is rejected.

In Excel: An F-test is carried out using a method that is also known as a single-factor analysis of variance (ANOVA) (see Chapter 7). Write the observations for the different populations into k adjacent columns of an Excel spreadsheet. Click TOOLS → DATA ANALYSIS → ANOVA: SINGLE FACTOR. Highlight all columns containing the data (check the COLUMNS button in the menu). Once the analysis is complete, the F-test statistic in the ANOVA table produced by Excel can be found in the column labeled "F." The corresponding p-value appears in the "P-value" column of the same table.

Example 6.6

For examples on the application of the F-test, see Sections 7.2.1 and 7.2.2.

6.2.4 Tukey's Test and Scheffé's Test

Once an F-test provides evidence of a difference among the k population means, it is desirable to determine which population means actually differ. Two methods, named after their inventors Tukey and

Scheffé, can be used. Neither method is implemented in Excel, but both are typically available in many other statistics software packages (e.g., SAS, R, S$^+$, Minitab, and SPSS).

Tukey's method performs pairwise comparisons of the k means (comparing each population to every other population separately). This method is preferable when only pairwise comparisons of means are of interest. For each comparison, Tukey's method tests whether or not the two population means differ. This sounds similar to conducting many two-sample t-tests, but it is not the same. Because all possible comparisons of pairwise means are made, the number of pairwise comparisons is taken into account so that the *overall* chance of making an error (among all comparisons) is controlled by the significance level or the investigator's willingness to be incorrect.

Scheffé's method is designed to draw conclusions about linear combinations of group means. For example, after the F-test has concluded that three groups do not have equal means, one could ask whether the means of groups one and two are the same and whether the mean of group three is twice as large as those of groups one and two.

Example for Scheffé: $H_0 : \mu_1 = \mu_2$ and $\mu_3 = 2\mu_1 = 2\mu_2$.

6.2.5 χ^2-test: Goodness-of-Fit or Test of Independence

The χ^2-test is used in two different applications.

- A "goodness-of-fit" application tests whether data have been generated by a specified mechanism.
- A "test of independence" application tests whether two observed factors occur independently of each other.

In both cases, observations are collected on two categorical variables (Section 3.1). The data are most often organized in the form of a CONTINGENCY TABLE where the rows and columns are labeled with the values that the two categorical variables assume. The cells of the table total how many observations of each particular variable combination were observed in the sample.

Example 6.7

Suppose in a population of fruit flies the observed variables are eye color (red or brown) and gender (male and female). Eye color and gender are recorded for each fly. The resulting contingency table has two rows and two columns. The cell corresponding to red eye color and male gender contains the number of male fruit flies with red eyes that were observed in the vial.

Goodness-of-fit test

Consider the case where observations on two categorical variables have been recorded in a contingency table. A model has been formulated for the occurrence of different factor combinations and it is of interest to determine whether or not the observed data conform to the hypothesized model (e.g., segregation of genes).

TEST STATISTIC: The goodness-of-fit χ^2-test statistic compares the observed counts to counts that one expects to see if the hypothesized (or expected) model is indeed correct.

$$\chi^2 = \sum \frac{(\text{observed count} - \text{expected count})^2}{\text{expected counts}} \sim \chi^2 \, (df = (r-1)(c-1))$$

Here, r stands for the number of rows and c stands for the number of columns in the contingency table.

ASSUMPTION:

- The distribution of the test statistic is approximately χ^2 distributed if the expected counts are large enough. Use this test only if the expected count for *each* cell is ≥ 5.

For 2×2 tables, Fisher's exact test provides a non-parametric alternative (i.e., no distribution assumptions are placed on the data) for the case that the sample sizes are small (see Section 6.3.2).

> **In Excel:** Write the observed counts in table form. You will have to compute the expected counts yourself by applying the hypothesized model to the observations and write them in another table of the same dimension. Then click in any empty cell and type "=CHITEST(TABLE1, TABLE2)" highlighting the observed and expected table, respectively. This command will return the p-value for the goodness-of-fit test.

Example 6.8

In dihybrid corn, different combinations of genes may lead to the following phenotypes: purple/smooth, purple/shrunken, yellow/

smooth, and yellow/shrunken. The four phenotypes are produced by two pairs of heterozygous genes, located on two pairs of homologous chromosomes. According to Mendelian genetics, one would expect the following ratio of phenotypes: 9:3:3:1 (i.e., the hypothesized model).

For one ear of corn with 381 kernels, each kernel has been classified as either smooth or shrunken and as either purple or yellow. The results are listed in the contingency table below:

Observed	Smooth	Shrunken	Total
Purple	216	79	
Yellow	65	21	
Total			381

1. Use significance level $\alpha = 0.05$.
2. Under the Mendelian model, nine out of 16 of the kernels are expected to be smooth and purple. To compute the expected counts for the Mendelian model, calculate 9/16, 3/16, 3/16, and 1/16 of the total number of observations ($n = 381$) and enter them in a new table:

Expected	Smooth	Shrunken	Total
Purple	214.3125	71.4375	
Yellow	71.4375	23.8125	
Total			381

3. and 4. The Excel command "=CHITEST(TABLE1, TABLE2)" applied to both the observed and expected count tables, respectively, yields a p-value of 0.189.

5. This number is large (compared to significance level $\alpha = 0.05$) and hence, the data do not supply evidence to reject the hypothesized Mendelian model.

Note that the test assumptions are satisfied, since the counts in the expected table are all ≥ 5.

χ^2-test for independence

When working with one population, it may be of interest to ask whether two factors are independent. Consider observations on two categorical factors that have been recorded in a contingency table.

HYPOTHESES:

$$H_0 : \text{the factors are independent}$$
$$H_a : \text{the factors are not independent}$$

TEST STATISTIC: If the two observed factors are indeed independent, then expected counts for each factor combination can be obtained by considering count totals and multiplying row and column totals of a contingency table.

$$\text{expected count} = \frac{\text{row total} \cdot \text{column total}}{\text{population total}}$$

The test statistic for independence using a χ^2-test is the same as the one for the goodness-of-fit test.

$$\chi^2 = \sum \frac{(\text{observed count} - \text{expected count})^2}{\text{expected counts}} \sim \chi^2 \; (df = (r-1)(c-1))$$

Here again, r stands for the number of rows and c stands for the number of columns in the contingency table.

> **In Excel:** Write the observed counts in table form. Compute all row and column totals. You will have to compute the expected counts yourself and write them in another table of the same dimension. Then click in any empty cell and type "=CHITEST(TABLE1, TABLE2)," highlighting the observed and expected table, respectively. This command will return the p-value for the χ^2-test for independence.

Example 6.9

In a vial of 120 fruit flies, an experimenter counts 55 male (and thus 65 female) flies. 24 of the male and 38 of the female flies have red eyes. All other flies have brown eyes. The question under investigation is whether gender and eye color are genetically linked.

The contingency table for these data is:

Observed	Red	Brown	Total
Male	24	31	55
Female	38	27	65
Total	62	58	120

1. Use significance level $\alpha = 0.05$.
2. The null hypothesis is that the traits for gender and eye color are unlinked or independent versus the alternative hypothesis that the traits are linked. If the traits for gender and eye color are independent, we expect $55/120$ flies to be males and $62/120$ flies to be red eyed. Thus, we expect

$$\left(\frac{55}{120}\right)\left(\frac{62}{120}\right)(120) = \frac{55 \cdot 62}{120} = 28.417$$

flies in the vial to be red-eyed males. Computing the remaining three expected counts yields the expected table:

Expected	Red	Brown	Total
Male	28.417	26.583	
Female	33.583	31.417	
Total			120

3. and 4. The "=CHITEST(TABLE1, TABLE2)" command in Excel applied to the observed and expected tables (do not highlight the totals, only the four cells containing the counts) yields a p-value of 0.105.
5. Since this value is large (compared to $\alpha = 0.05$), we fail to reject the null hypothesis and conclude that gender and eye color are independent or genetically unlinked.

6.2.6 Likelihood Ratio Test

A likelihood ratio test compares two different statistical models to each other for the purpose of deciding which of the two models better describes the given set of observations. A likelihood is the chance or probability of observing a given set of observations under a specified probability model.

Example 6.10

Suppose data are generated by crossing two F_1 hybrids and phenotyping the F_2 offspring. Among 40 F_2 progeny, 25 appear to be wild type (wt) and 15 mutant. The investigator compares two inheritance models. In one model, she would expect a 3:1 ratio of wild-type to mutant phenotypes, and in the other model, she would expect a 1:1 ratio of phenotypes. A choice is made by considering the likelihoods of both models given the information provided by the data. In both cases, the likelihood can be computed by using a binomial probability model (Section 3.4.1).

Model 1: $P(25 \text{ wt and } 15 \text{ mutant}) = \binom{40}{25}(.75)^{25}(.25)^{15} = 0.0282$

Model 2: $P(25 \text{ wt and } 15 \text{ mutant}) = \binom{40}{25}(.5)^{25}(.5)^{15} = 0.0366$

Both of these models have the same structure, but they differ in the value of the parameter $p = $ proportion of wild type. A direct comparison of the likelihoods concludes that it is more likely that the inheritance model is 1:1 (since $0.0363 > 0.0282$).

In many cases, one of the two models to be compared is more complex (i.e., has more parameters) than the other and the simpler model is a subset of the complex one. The models should have the same general structure and differ only through additional parameters in the more complex model. The model which has more parameters will always have higher likelihood because it has improved explanatory power over the simpler model. However, there is a trade-off between simplicity of the model and increasing likelihood. Specifically, simpler models are often easier to explain in a biological background.

TEST STATISTIC: If the sample size is large ($n > 30$), then the distribution of the likelihood ratio test (LRT) statistic is approximately a χ^2 distribution.

$$LRT = -2 \ln \frac{L_0}{L_a} \sim \chi^2(df)$$

Here, L_0 is the likelihood of the observations under the null hypothesis and L_a is the likelihood of the observations under the alternative hypothesis. The degrees of freedom of the χ^2-distribution of the test statistic is the difference in parameter number between the two models.

LOD SCORES: When assessing genetic linkage, results are sometimes formulated in the form of log-odds (LOD) scores. Similar to the LRT statistic, the LOD score compares the likelihoods L_0 of a null hypothesis model (no linkage) with an alternative model L_a (linkage).

$$\text{LOD} = \log_{10} \frac{L_a}{L_0}$$

In general, LOD scores are easier to interpret than LRT statistics. For instance, a LOD score of 2 means that the linkage model is 100 times more likely than the no-linkage model given the information provided in the data.

Note: A LOD score and the LRT statistic for the same null hypothesis and alternative are closely related. In fact,

$$LOD = \left(\frac{1}{2 \cdot \ln(10)} \right) LRT \approx 0.217 LRT$$

6.3   Non-parametric Tests

Many statistical tests are based on assumptions that are placed on the population from which the sample or data are drawn. For instance, the test statistic function used for a two-sample t-test (Section 6.2.1) only has a t-distribution if the data from both populations are normally distributed. If this is not the case, then the sample sizes need to be large so that the sample means will have at least approximate normal distributions according to the Central Limit Theorem (Section 3.5).

In many biological applications, the behavior or distribution of observations may be unknown since there may be little prior knowledge about the populations in an experiment. Additionally, sample sizes are often not large enough for the Central Limit Theorem to be effective.

NON-PARAMETRIC TEST: A testing procedure that does not make any distribution assumptions about the population from which the data are taken is called NON-PARAMETRIC. Non-parametric tests still make assumptions (such as equal variance assumptions, for instance), but they do not require knowledge of the distribution of the population.

Non-parametric alternatives exist for most of the commonly used statistical testing scenarios that require population distribution assumptions. Although the main advantage of the non-parametric methods is that their assumptions are less restrictive, the downside is that they are more tedious to compute when compared with conventional (parametric) procedures. If the data *do* satisfy the distribution assumptions of the conventional procedures, then the conventional procedures have more statistical power (Section 6.1.3) than their non-parametric counterparts. If the sample size is large, then there is usually little difference in the statistical power between the parametric and corresponding non-parametric procedures.

6.3.1   Wilcoxon-Mann-Whitney Rank Sum Test

This non-parametric alternative to the two-independent sample t-test can be employed to decide whether two population means are equal based on samples from both populations.

TEST STATISTIC: Suppose that the samples drawn from the two populations have sizes n_1 and n_2, respectively. The observations in the samples are ranked regardless of the population to which they belong, and the ranks for each population are added. Let R_1 denote the rank sum of observations from population one (you could also use population two, it does not matter).

$$U = n_1 n_2 + \frac{n_1(n_1+1)}{2} - R_1 \sim N\left(\mu = \frac{n_1 n_2}{2}, \sigma = \sqrt{\frac{n_1 n_2 (n_1 + n_2 + 1)}{12}}\right)$$

In Excel: There is no single command for a WILCOXON-MANN-WHITNEY test in Excel. However, the test statistic can be computed in a relatively simple stepwise manner.

1. Write the observations from both populations in *one column* of an Excel spreadsheet. In an adjacent column, keep track of which population each observation belongs to by recording the population number (1 or 2) in the column.
2. Using the DATA → SORT feature, sort the observations by magnitude (smallest first). Highlight the observation column, click DATA, and then click SORT. Activating the "Expand the selection" option will allow you to simultaneously sort the adjacent column (i.e., population labels).
3. Create another column with the ranks of the observations: Give the smallest observation rank 1, the next smallest rank 2, and so on. The largest observation should have rank $n_1 + n_2$. Break potential ties by averaging the ranks for repeated observations.
4. Add up all the ranks of observations in sample 1. To do this, sort all three columns by the population label column and then add the appropriate ranks using Excel's "=SUM()" feature.

continued

5. Compute the test statistic value U as:

$$U = n_1 n_2 + \frac{n_1(n_1 + 1)}{2} - R_1$$

Here, R_1 is the rank sum for population 1—it does not matter which one of your two populations you call population 1 and which one you call population 2; just be consistent.

6. This test statistic has approximately a normal distribution with a mean μ and standard deviation σ that can be computed using the sample sizes, n_1 and n_2.

$$U \sim \text{Normal} \left(\mu = \frac{n_1 n_2}{2}, \sigma = \sqrt{\frac{n_1 n_2 (n_1 + n_2 + 1)}{12}} \right)$$

7. A two-sided p-value for the Wilcoxon-Mann-Whitney test can be obtained in Excel by typing "=NORMDIST($-$ABS($U-\mu$), μ, σ, TRUE)." Here, U is the value of the test statistic that you computed in Step 5, and μ and σ are the mean and standard deviation of the normal distribution that you computed in Step 6.

ASSUMPTIONS:

- This test is non-parametric. It makes no assumptions on the distributions of the populations from which the samples are drawn.

- However, to be good representatives of their respective populations, both samples must be unbiased and have reasonably large sample sizes.

Example 6.11

Two strains of *Drosophila*, *D. melanogaster* and *D. simulans*, are compared with respect to the number of mature eggs in female ovaries for the purpose of studying reproductive efforts. Four females from each strain are dissected and mature eggs are counted. The results are:

| *D. melanogaster*: | 6 | 5 | 6 | 7 |
| *D. simulans*: | 8 | 3 | 5 | 4 |

We want to decide whether the number of mature eggs carried by females of the two species are on average the same or not.

1. Use significance level $\alpha = 0.05$.
2. The null hypothesis is that the species do not differ in number of eggs (means and variances are the same) with the two-sided alternative that the species differ.
3. The observed egg counts from each sample are recorded in one row. Population label (*melanogaster* $= 1$ and *simulans* $= 2$) are recorded in the adjacent row. The egg counts are ranked smallest to largest while accounting for the population label. Ties between repeated observations of the same magnitude are broken by averaging the ranks.

Observation	3	4	5	5	6	6	7	8
Population	2	2	1	2	1	1	1	2
Rank	1	2	**3.5**	**3.5**	**5.5**	**5.5**	**7**	**8**

The rank sum for population 1 (*melanogaster*) is $R_1 = 21.5$. Compute the U-test statistic as

$$U = (4)(4) + \frac{(4)(5)}{2} - 21.5 = 4.5$$

4. The test statistic has a normal distribution with mean

$$\mu = \frac{(4)(4)}{2} = 8$$

and standard deviation

$$\sigma = \sqrt{\frac{(4)(4)(9)}{12}} = \sqrt{12} = 3.464.$$

The p-value for this two-sided test can be computed in Excel as "=NORMDIST($-$ABS(4.5$-$8), 8, 3.464, TRUE)" which yields a p-value of 0.000450.

5. This value is smaller than the significance level α. Hence we can reject the null hypothesis and conclude that the average egg number differs in the two strains.

6.3.2   Fisher's Exact Test

Fisher's exact test is a non-parametric test for independence. It accomplishes the same thing as the χ^2-test for independence in the special case when sample sizes are small and the contingency table has dimension 2×2. The χ^2-test for independence itself does not make any distribution assumptions, and is thus a non-parametric test. Since the χ^2-test requires large sample sizes, Fisher's exact test may be used as a (less powerful) alternative when sample sizes are small.

For a 2×2 contingency table with possibly low observed and/or expected counts, we want to decide whether one treatment is preferable to another. Note that this is equivalent to asking whether there is an association between treatment and outcome.

	Treatment 1	Treatment 2	Row total
Outcome 1	a	b	$a + b$
Outcome 2	c	d	$c + d$
Column total	$a + c$	$b + d$	$n = a + b + c + d$

TEST STATISTIC: Using a combinatorial approach, we compute the probability of obtaining more extreme data by chance than the data observed if the row and column totals are held fixed. For example, the probability of randomly observing the *same* outcome as the one in the table is

$$\frac{\binom{a+b}{a}\binom{c+d}{c}}{\binom{n}{a+c}}$$

Here, $\binom{a+b}{a}$ for instance, is the binomial coefficient introduced in Section 3.4.1. If we shift the observations in the table one by one so that they more strongly support the alternative, while holding the row and column totals fixed, we create a series of new tables. For each shift in observations, and each new table, the probability of obtaining more extreme data than by chance can be calculated. If the probabilities for each table are added, this number is the p-value of the test. If there are too many representations of the data to consider, an alternative approach to Fisher's exact test is the permutation test.

ASSUMPTIONS:

- Both samples must represent the populations from which they have been drawn (no bias).
- There are no assumptions on distributions or sample sizes for this test. However, if the sample sizes are very small, then the power of the test will be very low.

Example 6.12

Root rot is a common disease that befalls trees in the United States. A fungus infects the trees and rots their roots, eventually killing the tree. In a small urban park, researchers counted infected and healthy trees. The park has 12 plane trees, of which 11 were found to be infected, and 8 ash trees, of which only two were infected. Clearly, plane trees tend to be more infected than ash trees. But can the relationship between root rot and tree type be quantified with a p-value?

	Infected	Healthy	Total
Plane	11	1	12
Ash	2	6	8
Total	13	7	20

The null hypothesis of Fisher's exact test is that root rot and tree species are independent. It is possible that of the 13 infected trees in the park, 11 (or more) were plane trees, just by chance, even if the fungus befalls both tree species with equal probability. In fact, the probability that the infection pattern of the trees is exactly as observed is

$$\frac{\binom{12}{11}\binom{8}{2}}{\binom{20}{13}} = 0.00433$$

A p-value is the probability of observing extreme data by chance if the null hypothesis were true. More extreme data in this example would be even more infected plane trees and fewer infected ash trees.

	Infected	Healthy	Total
Plane	12	0	12
Ash	1	7	8
Total	13	7	20

The probability of this outcome is

$$\frac{\binom{12}{12}\binom{8}{1}}{\binom{20}{13}} = 0.000103$$

It is not possible to make the data even more extreme while keeping the row and column totals fixed. Thus, the p-value for the null hypothesis of no association is

$$p = 0.00433 + 0.000103 = 0.00443$$

Since this value is small ($< \alpha = 0.05$), we reject the null hypothesis and conclude that there is an association between fungus infection and tree species.

6.3.3   Permutation Tests

If the null hypothesis represents the situation where there is absolutely no difference (in mean, variance, or distribution) between two populations, then observations from the two populations are entirely interchangeable. Without making assumptions on what the distributions are (only assuming that they are the *same*), the distribution of a test statistic can be obtained by considering many different permutations of the observed data and computing a test statistic for each permuted data set. Permuting the observations can also be thought of as randomly reassigning group identification labels. For example, for sample size n there are n labels or indices. If all labels or indices were placed in a hat, drawn without replacement, and assigned at random to the observations, this would be one permutation of the data. One can imagine producing many permuted data sets by repeating this process.

Theoretically, any function that represents the difference in sample means can be considered for this purpose. However, conventional test statistic functions (e.g., the two-sample t-test test statistic) are typically employed. For complicated test statistics and large sample sizes, computations can quickly become very time intensive (millions of possible permutations must be considered). If the number of possible permutations becomes excessive (there are already 184,756 ways to permute 10 treatments and 10 controls), a random subset of possible permutations may be considered instead. Statisticians call this method MONTE CARLO. The only restriction on permutation tests is that under the null hypothesis the distributions (both in location and shape) of the populations must be equal. If two populations have unequal variances, for instance, the individuals cannot be randomly permuted.

Once the permuted data sets are created and a test statistic computed for each data set, the resulting test statistic values are ordered (or ranked) from smallest to largest, and the $(1 - \alpha)\%$ cutoff value is identified as the $(1 - \alpha)^{th}$ percentile of the ordered permutation test statistic values. The test statistic value as calculated from the original data is compared to this cutoff value. If the test statistic from the original data is greater than the cutoff value obtained from permuting the data, then the null hypothesis of no association (or randomness) is rejected in favor of the alternative hypothesis.

Example 6.13

Cell cultures are grown both in standard medium and in medium supplemented with α-tocopherol (vitamin E). For each plate, the number of cell divisions is counted after a pre-determined time. To investigate whether the average growth is the same under both conditions, a permutation test is performed. The data consist of six observations on standard medium and four observations on medium supplemented with vitamin E.

Standard medium	45	110	63	55	67	75
Vitamin E medium	100	60	72	51		

Since the null hypothesis is that the means for both treatments are the same and it is reasonable to assume the distributions to have the same shape and same variance, a permutation test may be performed. There are $\binom{10}{6} = 210$ different ways to reorder the ten observations into two groups of size 4 and 6, respectively, and thus 210 resulting test statistics that represent the distribution of the test statistic under the null hypothesis.

	Group 1						Group 2				$\bar{x}_1$	$\bar{x}_2$
Original	45	110	63	55	67	75	100	60	72	51	69.167	70.75
Permutation 1	75	67	55	72	45	63	110	60	100	51	62.833	80.25
Permutation 2	72	55	60	51	45	110	63	75	67	100	65.5	76.25
Permutation 3	60	45	63	55	67	51	110	100	72	75	58.833	89.25
$\vdots$												
Permutation 210	72	110	100	67	63	45	75	51	55	60	76.167	60.25

To compare the means of the two groups, we use the test statistic function from the two-independent-sample t-test with equal variance assumption.

$$\frac{\bar{x}_1 - \bar{x}_2}{\sqrt{\frac{(n_1-1)s_1^2+(n_2-1)s_2^2}{n_1+n_2-2} \cdot \left(\frac{1}{n_1} + \frac{1}{n_2}\right)}}$$

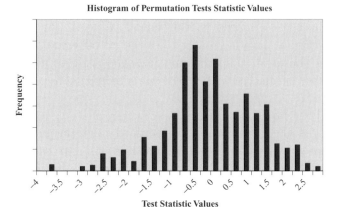

FIGURE 24. For each permutation of observations, the group means and standard deviations are recomputed. Based on these statistics, the test statistic value is recomputed for each permutation. The permutation test statistics are graphed in the form of a histogram. The p-value of the test is the percentage of test statistics more extreme than the one observed in the original data.

Unlike in a t-test, however, we make no prior assumptions on the distribution of this test statistic. Rather, we compute the 210 test statistic values arising from all possible permutations of the data and use *their* distribution (Fig. 24) to derive a p-value for the original set of observations.

For the original set of observations, the value of the test statistic used here is -0.111. The p-value for the permutation test is the percentage of permuted samples for which the test statistic value is more extreme (smaller than -0.111 or larger than $+0.111$). Since this number ($p = 0.908$) is large, we fail to reject the null hypothesis that the two treatment means are equal.

Note: If the number of possible permutations is very large, then it is not necessary to actually compute all possible permutations. Instead, it is sufficient to compute a large (≥ 1000) number of *random* permutations or randomly select a large number of all possible permutations. This method is commonly referred to as the MONTE CARLO method.

6.4 E-values

Some biological applications, most prominently BLAST, return a quantity called an E-VALUE in addition to, or instead of, a p-value.

What is the difference in meaning between an E-value and a p-value? "E" in this case stands for "expect" and represents the average number of hits in a given database with a score bigger than or equal to that of the query. The null hypothesis in a BLAST search is that the query is a random (DNA) sequence with no match in the database. A score is computed for every pairwise comparison between the query and candidate sequences from the database. Two sequences with a high degree of similarity will receive a high score.

A BLAST search returns the top matching sequences from the database together with an E-value for each match. The E-value counts how many sequences in the database one would expect to see receive scores exceeding the observed score in a random matching. This quantity also takes the size of the database into account. It is easier to interpret the difference between E-values of 5 and 10 (5 or 10 expected random matches in the database) than to interpret the difference between p-values of 0.99326 and 0.99995. Large E-values mean that one would expect to find many random matches in the database and that it is hence doubtful that the result is a true match.

Note: E-value and p-value are related through the equation

$$p = 1 - e^{-E}.$$

For very small values ($E < 0.01$), E- or p-values are virtually identical.

7   Regression and ANOVA

In many experiments, several variables are observed simultaneously with the goal of learning more about their relationships. Some of these variables may be specified by the experimenter (e.g., treatment and control conditions or choice of organisms involved in the experiment) and some may be observed as reactions. Generally, variables that can be varied by the experimenter (e.g., strain of organism, growth conditions, treatments, etc.) are referred to as EXPLANATORY or PREDICTOR variables. Variables that are measured or observed as reactions are referred to as RESPONSE variables if they are of primary interest or as explanatory variables if they are measured but are not of primary interest in the experiment.

Depending on the number and type (categorical or quantitative) of variables, different statistical models are fit to the observations to draw conclusions about the experiment or make predictions for future experiments. Generally, models with one quantitative response variable are called UNIVARIATE and models with two or more response variables are called MULTIVARIATE. Multivariate models are considerably more complicated than univariate models because possible associations between response variables must be taken into account. Up to this point, and throughout the rest of this Manual, we focus only on univariate models in which one response variable is observed together with one or more predictor variables.

Regression and ANOVA are two of the most commonly used statistical techniques. They have a lot in common (Section 7.3) and differ primarily in the types of variables observed. In a regression model, a quantitative response variable is written as a linear function of one or more quantitative predictor variables plus an *independent error term*. Statistical inference includes estimating the values of the model parameters, providing confidence intervals, and conducting hypothesis tests. The model will make assumptions on the distribution of the error terms that need to be verified before any conclusions are drawn. Data from a regression analysis are most commonly displayed in scatter plots (Fig. 25, left).

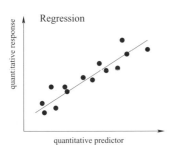 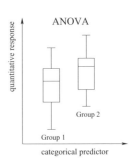

FIGURE 25. (Left) In a regression model, a quantitative response is explained through one or more quantitative predictors. (Right) In an ANOVA model, a quantitative response is explained through one or more categorical predictors.

If the response variable is quantitative but the predictor variables are categorical, an ANOVA MODEL is used. In this setup, the main question is whether or not the levels of the categorical predictor variables have significant influence on the average response. To answer this question, the variation among group means is compared to the variation within the groups defined by the categorical predictor(s) (Fig. 25, right).

Model	Response	Predictor(s)
one-way ANOVA model	one quantitative	one categorical
two-way ANOVA model	one quantitative	two categorical
simple regression	one quantitative	one quantitative
multiple regression	one quantitative	two (or more) quantitative
logistic regression	one categorical	one (or more) quantitative

If all variables in an experiment are categorical, statistical inference is usually focused on evaluating associations between variables via χ^2-tests (Section 6.2.5).

7.1   Regression

The simplest statistical model is the one in which one quantitative explanatory variable is used to predict one quantitative response variable. This is called a simple linear regression model. Recall from Section 3.8 that the correlation coefficient r may be used as a statistical measurement for the strength of association between two quantitative variables. If the association is strong, then in a scatter plot of

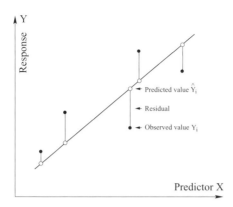

FIGURE 26. In least squares regression, the line is fit to minimize the sum of squared residuals.

the two variables, the data points may be approximated by a straight line. Conventionally, response variables are plotted on the y-axis and explanatory variables on the x-axis (Fig. 26).

Which straight line approximates the data points best? There are several ways to make this decision. By far the most popular method (and the one implemented in Excel) is LEAST SQUARES regression. The line is chosen so that the squared vertical distances between data points and the line are minimized. The distances between observations and their predicted counterparts are called RESIDUALS (Fig. 26). For a regression model to be considered good, the residuals should be normally distributed over the predictor variable range with roughly constant standard deviation σ.

Regression line equation: The mathematical equation for a line is

$$y = b_0 + b_1 x.$$

Here, y and x are the response and predictor variables, respectively, b_0 is the INTERCEPT, and b_1 is the SLOPE of the line.

Due to experimental error, most observations will not lie directly on the line. Therefore, the observations y_i are modeled as observations from a normal distribution with mean μ_y that depends on x and standard deviation σ. The residual terms ϵ_i represent the variation in the observations y around the mean μ_y. They are assumed to be normally distributed with mean 0 and constant standard deviation σ.

Population: $y_i \sim \text{Normal}(\mu_y, \sigma^2)$,   where $\mu_y = \beta_0 + \beta_1 x$
Sample: $y_i = b_0 + b_1 x_i + \epsilon_i$

Estimates for the intercept b_0, the slope b_1, and the residual standard deviation $\hat{\sigma}$ can be computed (in Excel) by fitting a least squares regression line to the observations.

In Excel: Write the observations for your predictor and response variables into two columns of a worksheet. Observations on the same individual need to appear in the same row. Click TOOLS → DATA ANALYSIS → REGRESSION. Highlight your response (y) variable and then your predictor variable (x). Excel will provide a regression line if you select LINE FIT PLOTS. To check the fit of your model (recommended!), you can select RESIDUAL PLOTS and NORMAL PROBABILITY PLOTS to check whether the residuals are normally distributed with constant variance (Section 7.1.6).

Correlation and regression

The closer the correlation coefficient of two sets of quantitative measurements is to 1 or -1, the stronger the association between the two variables. In terms of regression, strong association means that the data points of the two measurements lie close to their least squares regression line. Statistically, the square of the correlation coefficient, r^2, is the portion of the variation in the response variable y that is explained by the predictors, x (or, explained by the regression of y on x).

In Excel: When a regression analysis is performed in Excel, the output is organized in the form of three tables. The "Regression Statistics" table contains information about the correlation coefficient and the percentage of explained variation. Here, MULTIPLE R is the correlation coefficient r of the predictor and response variable. R SQUARE is the square of the correlation coefficient—this number represents the percentage of variation of the response explained through the predictor. High values of r^2 (close to 1) mean that the predictor can predict the response very well. Adjusted R Square is a quantity that is only relevant for multiple linear

continued

regression (Section 7.1.4). The STANDARD ERROR σ is the estimate of the residual standard deviation, and OBSERVATIONS is the number of data pairs that were used for the analysis.

SUMMARY OUTPUT	
Regression Statistics	
Multiple R	correlation
R Square	square of correlation
Adjusted R Square	R^2 adjusted for predict or number
Standard Error	estimate of error variance
Observations	n

7.1.1 Parameter Estimation

The values b_0, b_1, and $\hat{\sigma}$ of the model parameters computed by Excel obviously depend on the data. If the experiment were to be repeated under similar conditions, the observations could not be expected to be exactly the same. Different measurement errors lead to (slightly) different estimates. In this sense, the model parameters are random variables and we can ask statistical questions about them. In particular, we can formulate confidence intervals for the parameters or conduct hypothesis tests for them.

In Excel: When a regression analysis is performed in Excel, the estimates of the model parameters are computed and reported in a table. The coefficients table contains the estimates b_0 of the intercept and b_1 of the slope. In the same table, in the columns to the right are upper and lower limits for a (usually) 95% confidence interval for the parameters. The estimate of the residual standard deviation $\hat{\sigma}$ can be read from the "Regression Statistics" table in the Standard Error line. The MSE (mean squared error) cell in the ANOVA table will report the estimated residual variance $\hat{\sigma}^2$.

ANOVA					
	df	*SS*	*MS*	*F*	*Significance F*
Regression	1	SSM	MSM	F-statistic value	p-value
Residual	n-2	SSE	MSE		
Total	n-1				

	Coefficients	*Standard Error*	*t Stat*	*P-value*	*Lower 95%*	*Upper 95%*
Intercept	b0	SE_{b0}	t-statistic value	p-value	95% confidence interval for b0	
X Variable 1	b_1	SE_{b1}	t-statistic value	p-value	95% confidence interval for b1	

How should the slope estimate b_1 computed in a simple linear regression analysis be interpreted? For each unit increase in the predictor variable x, the response y increases on average by b_1 units. If the measurement units of one or more variables in the regression model are changed (e.g., from inches to cm), this will influence the values of all regression parameters and the analysis should be repeated to obtain new parameter estimates.

7.1.2   Hypothesis Testing

The most meaningful question to ask of the model parameters in a regression model is whether or not they are really necessary. A parameter does not need to be included in the model if its value is zero. The p-values in the coefficients table returned by Excel correspond to this null hypothesis. If a p-value is small, its corresponding slope is significantly different from zero. Thus, the corresponding predictor has a significant influence on the response. On the other hand, if a slope p-value is large, then the slope is (almost) zero. In this case, the predictor has very little influence on the response. Predictors with large p-values for their slopes may potentially be excluded from a regression model (Section 7.1.5).

Example 7.1

Researchers at the University of California in Davis have studied the link between flavonoids in organically and conventionally grown tomatoes and the amount of nitrogen in the fertilizer used on the plants. Organic fertilizer, such as manure, contains considerably less nitrogen than commercially produced fertilizers. The researchers measured the nitrogen application (in kg ha/yr) as well as the flavonoid compound Kaempferol (in mg/g) over the course of $n = 6$ years. For comparison, tomatoes were grown on adjacent plots in the same environmental conditions.

Nitrogen	Kaempferol
446	43
347	48
278	56
254	67
248	90
232	86

SUMMARY OUTPUT

Regression Statistics	
Multiple R	0.82683143
R Square	0.68365021
Adjusted R Square	0.60456276
Standard Error	12.3354874
Observations	6

ANOVA

	df	SS	MS	F	Significance F
Regression	1	1315.34301	1315.34301	8.64423161	0.0423846
Residual	4	608.656994	152.164248		
Total	5	1924			

	Coefficients	Standard Error	t Stat	P-value	Lower 95%	Upper 95%
Intercept	124.656117	20.906058	5.96267916	0.0039723	66.611594	182.700639
Nitrogen	-0.1983029	0.0674475	-2.9401074	0.0423846	-0.3855671	-0.0110386

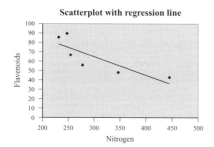

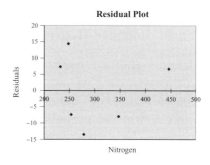

FIGURE 27. Excel output produced for the simple linear regression example. The output includes the summary table, ANOVA table and coefficients table as well as a line fit plot and the residual plot.

Using Excel to fit a simple linear regression model with nitrogen as the predictor (x) variable and Kaempferol as the response (y) results in the output shown in Figure 27.

To interpret the output, we can say that the nitrogen level in the soil is a significant $(p = 0.042)$ predictor for flavonoid content in the tomato fruit. In fact, 82.68% of the variation in flavonoid is explained by nitrogen. We can see that the flavonoid content *decreases* when more nitrogen is applied to the soil (in commercially grown tomatoes), since the regression coefficient for nitrogen is negative $(b_1 = -0.198)$. This means that for every kilogram of nitrogen added to a hectare of soil, the flavonoid content decreases on average by 0.19 mg per g of tomato.

In this example, it makes no sense to also interpret the intercept parameter in the context of the problem. Theoretically, $b_0 = 123.65$ is the amount of flavonoid contained in plants grown entirely without nitrogen. However, plants would not grow at all under these conditions which makes interpretation of this number meaningless.

To check whether the assumptions for the statistical model are satisfied, we have to take a look at the residuals. It is hard to discern a pattern in the residual plot, since the number of observations ($n = 6$) is rather small. A PP-plot of the residuals (not shown here) may be used to check the residuals for normality. Again, since the number of observations is small, it will be difficult to discern a pattern in the residual plot.

7.1.3 Logistic Regression

Consider the case where a bivariate categorical response (e.g., subject lives or dies, or experiment succeeds or fails) is to be predicted from one (or more) quantitative predictors. In this case, we are most interested in the probability p of success (subject lives, experiment succeeds) for a given level of the predictor x. We cannot directly fit a linear regression model since the response variable takes on only two values. Instead, consider the *probability* that the response is a success if the experiment is conducted at a particular predictor level. Fit a linear regression model to the logit function of this probability:

$$\text{logit}(p_i) = \ln\left(\frac{p_i}{1 - p_i}\right) = b_0 + b_1 x_i$$

Here, p_i is the proportion of observed successes if the experiment was conducted at predictor level i. Let x_i denote the predictor value for individual or trial i and make $p_i = 1$ for a success and $p_i = 0$ for a failure.

Example 7.2

Excel is not equipped to conduct a logistic regression analysis, but many other software programs (R, SPSS, Minitab, etc.) are. The output will be in the form of the regression coefficients for the logit function b_0, b_1. The success probability can then be estimated by solving the model equation for p:

$$\ln\left(\frac{p}{1 - p}\right) = b_0 + b_1 x \quad \Leftrightarrow \quad p = \frac{e^{b_0 + b_1 x}}{1 + e^{b_0 + b_1 x}}$$

7.1.4 Multiple Linear Regression

Sometimes more than one predictor variable is used to explain the behavior of one quantitative response. For example, several genetic markers may be used to predict a quantitative trait locus (QTL). The regression model for a multiple regression model with k predictor variables $x_1, \ldots, x_k$ and response y becomes

Population: $y \sim \text{Normal}(\mu_y, \sigma^2)$, where $\mu_y = \beta_0 + \beta_1 x_1 + \cdots + \beta_k x_k$
Sample: $y_i = b_0 + b_1 x_{i1} + \cdots b_k x_{ik} + \epsilon_i$

As in the simple linear regression model, the response y is modeled as a normally distributed random variable with mean μ_y and constant variance σ^2. The mean μ_y is expressed as a linear combination of the predictor variables. There is still only one intercept b_0, but unlike the linear regression model, there are now k slopes $b_1, \ldots, b_k$, one corresponding to each predictor variable. The ϵ_i are the residuals, and the model assumes that they are normally distributed with mean zero and constant standard deviation σ.

NOTATION: For the multiple linear regression model, notation is

y_i i^{th} observation on the response
$\hat{y}_i$ what we expect the ith observation to be, given its predictors.
x_{ij} i^{th} observation of predictor variable j

In Excel: A multiple linear regression analysis is very similar to the simple linear regression case. Write the observations on your predictors and response into columns of a worksheet. Label each column with an appropriate name. Click TOOLS → DATA ANALYSIS → REGRESSION. Highlight the response y, and then all the predictor variables $(x_1, \ldots, x_k)$ respectively, including labels (check Labels box if appropriate). To check the assumptions of your model (recommended!), select RESIDUALS, RESIDUAL PLOTS, and NORMAL PROBABILITY PLOTS.

The output of a multiple linear regression analysis in Excel contains three tables (Fig. 28). The REGRESSION STATISTICS table reports the correlation coefficient for the multiple regression. Its square is the percent of variation in the response that can be explained by the predictors in the model. The STANDARD ERROR is the estimate of the residual standard deviation $\hat{\sigma}$, and n is the number of observations.

SUMMARY OUTPUT						
Regression Statistics						
Multiple R						
R Square	% variation					
Adjusted R Square						
Standard Error	$\hat{\sigma}$					
Observations	n					
ANOVA						
	df	*SS*	*MS*	*F*	*Significance F*	
Regression	k	SSM	MSM	F-test	p-value	
Residual	n-k-1	SSE	MSE			
Total	n-1	SST				
	Coefficients	*Standard Error*	*t Stat*	*P-value*	*Lower 95%*	*Upper 95%*
Intercept	b_0	$s(b_0)$	t-test	p-value	CI - lower	CI - upper
Predictor 1	b_1	$s(b_1)$	t-test	p-value	CI - lower	CI - upper
Predictor 2	b_2	$s(b_2)$	t-test	p-value	CI - lower	CI - upper
Predictor 3	b_3	$s(b_3)$	t-test	p-value	CI - lower	CI - upper

FIGURE 28. Schematic of the output produced by Excel after performing a multiple linear regression analysis with $k = 3$ predictors and n observations.

One (pessimistic) hypothesis test that one can conduct for a multiple regression analysis is whether or not any of the predictors have an effect on the response. This corresponds to testing the null hypothesis

$$H_0 : \beta_1 = \beta_2 = \cdots = \beta_k = 0 \text{ (all slopes equal to zero)}$$

This null hypothesis is tested against the alternative that at least one of the predictors has a non-zero slope. The test statistic and the corresponding p-value for this F-test can be found in the analysis of variation (ANOVA) table of the regression output. If the p-value in the SIGNIFICANCE F column of the ANOVA table is small, then it means that at least one of the k predictors in the model has an effect on the response. If this p-value is large (which rarely happens), then it tells you that your experiment was poorly designed, and none of the observed predictors has any significant influence on the response. The ANOVA table also contains the MSE (mean squared error), which is the estimated residual variance $\hat{\sigma}^2$.

The third table in the output (Fig. 28) contains the estimates of the individual slopes $b_0, b_1, \ldots, b_k$. It also reports 95% confidence intervals for those slopes and the p-values of t-tests that check each slope individually and ask whether or not it is equal to zero.

$$H_0 : \beta_j = 0$$

If a t-test for a slope has a small p-value, then this means that the corresponding predictor has a significant effect on the response and thus belongs in the regression model. If, on the other hand, the p-value is large, it means that the slope is (almost) zero and that the predictor could be excluded from the model.

7.1.5   Model Building in Regression—Which Variables to Use?

A good statistical model is as simple as possible while at the same time achieving high explanatory power. Simple in the context of regression means using as few predictor variables as possible. The explanatory power of the model can be measured by the percentage of variation in the response explained through the predictors (R^2). Including additional predictors in the model will always improve the explanatory power, but a small increase in R^2 may not be worth including another parameter.

The variable ADJUSTED R SQUARE takes the number of variables and the explanatory power of the model into account. It can be used as one tool in choosing a good regression model (pick the model with the highest adjusted R^2).

GENERAL STRATEGY: Building a good multiple regression model means selecting a minimal subset of predictors which has good predictory power for the response. To find the best regression model, begin with all predictor variables on which you have collected data. Fit a multiple linear regression model.

1. Look at R^2. The full model which has all predictors will have the highest R^2.
2. Look at the p-value for the ANOVA F-test. Proceed only if this p-value is small (less than 0.05).
3. Look at the p-values for the t-tests for the individual predictors. Variables with small p-values correspond to good predictors and those with large p-values correspond to not-so-good predictors.
4. Exclude the worst predictor (largest t-test p-value) and refit the model with the remaining predictors. Start over at 1.

Terminate this procedure when all predictors are significant (slopes have p-values less than 0.05) and when the exclusion of further predictors causes a large decrease in R^2. Don't forget to refit the model for your final choice of predictor variables. Every change in the set of predictors used will change the estimates for the slopes and the residuals. Check the model assumptions (normality and constant variance of residuals, Section 7.1.6) on the residuals computed for the final model.

7.1.6 Verification of Assumptions

Every statistical model relies on some kind of distribution assumptions. All inferences drawn from the analysis such as confidence intervals, hypothesis tests, and predictions for future experiments are *only* valid if those assumptions are (reasonably) satisfied. Computations using software can always be performed for a regression analysis, regardless of whether or not the assumptions are satisfied. It is the responsibility of the investigator to check whether or not the model is actually valid before using (or publishing) any conclusions.

> **Note:** In regression and ANOVA models, the assumption is that the residual terms ϵ_i are normally distributed with mean zero and constant variance σ^2.

Because of the way the model is formulated, the residual terms automatically have mean zero. This leaves two assumptions to check: Normality of the error terms and the constant variance assumption.

NORMALITY ASSUMPTION: A regression analysis in Excel will produce the values of the residuals if the RESIDUALS box is checked under the regression menu. The NORMAL PROBABILITY PLOT produced by Excel can be used as a rough guide for normality of the residuals. Preferably, produce a PP-plot or QQ-plot of the residuals instead (Section 3.4.3). In all three cases, normally distributed residuals should lie close to a line (45° line in the PP- or QQ-plot option). If there is a strong S-shape visible, then this is a sign for non-normally distributed residual terms.

CONSTANT VARIANCE ASSUMPTION: To check whether the residuals have approximately constant variance, use Excel to produce so-called RESIDUAL PLOTS. In these plots, the values of the residuals are plotted in a scatterplot against the predictor variables (or against the response variable). Excel will automatically produce residual plots

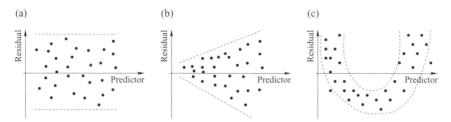

FIGURE 29. To check the constant variance assumption of the error terms produce a residual plot. If there is no pattern in the residual plot (a), the constant variance assumption is satisfied. A V-pattern (b) or a U-pattern (c) in a residual plot is an indication that the constant variance assumption is *not* satisfied.

against the predictor variables, if the option RESIDUAL PLOTS in the regression menu is checked. Look for any kind of pattern in the residual plot. V-shapes, U-shapes, or S-shapes are indicators that the residual variance is not constant (Fig. 29).

7.1.7   Outliers in Regression

Sometimes, a few observations do not fit into the overall regression scheme that the rest of the data follow. These points are called outliers (Fig. 30). The decision whether to call a point in a regression analysis an outlier is a subjective one. To make this decision, it is helpful to consider the residual plot (Section 7.1.6). If one residual falls outside of the pattern exhibited by the rest of the observations,

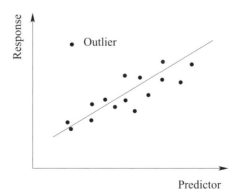

FIGURE 30. Data points that do not fit well into the regression model for the rest of the data are called outliers. These observations can be recognized by the unusual magnitude of their residuals.

then the observation corresponding to the residual may be an outlier.

To decide whether to exclude an outlier from the statistical data
analysis, several points need to be considered:

1. Is the outlier influential for the statistical model? This means
   that if the analysis is conducted twice—once for the whole
   data set and once for the reduced set in which the outlier is
   omitted—do the estimates of the model parameters (R^2, σ,
   slopes, p-values for slopes, etc.) change very much? If there is
   not much change, then it does not matter (much) whether the
   outliers are included or not. When in doubt, err on the side of
   caution and include every observation (even potential outliers).

2. Is there reason to believe that the observation does not follow
   the pattern because of some type of mistake made when the
   measurement was taken? Is it possible to repeat parts of the
   experiment to replace this observation (recommended)? If it
   can be established that some kind of technical error was made
   and the outlier is influential, the point may be excluded from
   analysis.

3. Is there reason to believe that some biological effect causes
   this observation to be very different from the others? If this is
   the case, then the point should *not* be excluded, even if it is
   influential. Instead, the model needs to be adjusted to incor-
   porate this biological effect.

7.1.8   A Case Study

A complete multiple regression analysis will likely consist of several
steps.

1. Fit a regression model with all available predictor variables.
2. Look at the residuals and verify the model assumptions. Transform variables if necessary.
3. Once the residuals are normally distributed, look at the predictor variables. If necessary, remove variables that are not
   significant for the response. Refit the model after each variable
   removal.

4. Choose a final model. Weigh the model fit (high R^2 and adjusted R^2) against simplicity of the model (simpler is better) and significance of predictor variables in the model (small p-values for the predictors).

5. Once more, verify the assumptions for this final model. If desired, draw plots showing the regression line. If you are concerned about outliers, identify potential outliers and repeat the above analysis for the data set with outliers removed.

Example 7.3

Fifty species of oak trees grow in the United States. Researchers studied the relationship between acorn size and the size of the geographic region in which the trees grew. They collected data (tree SPECIES, REGION of growth (0=Atlantic, 1=California), ACORN size, tree HEIGHT, and RANGE) from 39 species of oak. Acorn size is expressed as a volume computed from length and width measurements on the acorns. Parts of the data are shown in the following table:

SPECIES	REGION	RANGE	ACORN	HEIGHT
Quercus alba L.	Atlantic	24196	1.4	27
Quercus bicolor Willd.	Atlantic	7900	3.4	21
Quercus Laceyi Small.	Atlantic	233	0.3	11
$\vdots$				$\vdots$
Quercus vacc. Engelm.	California	223	0.4	1
Quercus tom. Engelm.	California	13	7.1	18
Quercus Garryana Hook.	California	1061	5.5	20

In this example, acorn size is the response, and range, height, and region are the predictors. Note that tree species is neither a predictor nor a response, as this variable identifies the individual trees measured in this experiment. The region variable is categorical (with values Atlantic and California) and it can be recoded by a numeric "dummy" variable (see Section 7.3). Specifically, Atlantic is coded as 0 and California is coded as 1.

Excel is used to fit a multiple linear regression with ACORN as the response and RANGE, HEIGHT, and REGION (coded by 0 and 1) as the predictors. A PP-plot of the residuals for this model shows that

the residuals are clearly not normally distributed, since both the
ACORN as well as the RANGE variable are strongly skewed to the
right.

To fix this problem, both the ACORN variable and the RANGE variable are natural log-transformed. That means that new variable
columns named LN(ACORN) and LN(RANGE) are created that contain the natural logarithms of the original ACORN and RANGE measurements. The transformed multivariate regression model that is
fitted next is of the form

$$\ln(\text{Acorn Size}) = \beta_0 + \beta_1 \ln(\text{Range}) + \beta_2 \text{Height} + \beta_3 \text{Region} + \epsilon$$

Excel output for the transformed model looks like this:

SUMMARY OUTPUT

Regression Statistics	
Multiple R	0.51637064
R Square	0.26663864
Adjusted R Square	0.20377909
Standard Error	0.86590347
Observations	39

ANOVA

	df	SS	MS	F	Significance F
Regression	3	9.5413992	3.1804664	4.24181622	0.01170792
Residual	35	26.2426089	0.74978883		
Total	38	35.7840081			

	Coefficients	Standard Error	t Stat	P-value	Lower 95%	Upper 95%
Intercept	-2.4583841	1.11119506	-2.2123785	0.03355945	-4.71423	-0.2025382
Height	0.02012296	0.01745603	1.15278011	0.2568133	-0.0153147	0.05556058
Region	1.41361301	0.48697975	2.9028168	0.0063613	0.42499156	2.40223447
ln(Range)	0.30614821	0.13044988	2.34686453	0.02471922	0.04132086	0.57097555

From the output, we can see that both REGION and the transformed
RANGE variable are significant for the transformed ACORN size. In
fact, researchers believe that larger seed size attracts larger animals
which are able to carry the seeds farther. The tree HEIGHT is not
significant for the ACORN size (p-value for HEIGHT is $p = 0.257 >
0.05$). Thus, this variable can be removed from the model, yielding
the transformed and reduced model:

$$\ln(\text{Acorn Size}) = \beta_0 + \beta_1 \ln(\text{Range}) + \beta_2 \text{Region} + \epsilon$$

With the HEIGHT variable removed from the model, the R^2 decreases
to $R^2 = 0.239$ (adjusted $R^2 = 0.197$). Recall that for the transformed
model that included HEIGHT, the $R^2 = 0.267$ (adjusted $R^2 = 0.204$).
The simplicity of the reduced model is preferable even with a smaller
R^2 value.

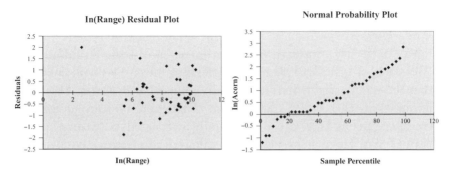

FIGURE 31. Residual plot and normal probability plot produced by Excel for the Acorn size example. There is one potential outlier (in red) visible in the residual plot. Otherwise, the residual plot does not show any pattern. The Probability plot demonstrates that the residuals in the reduced model are approximately normally distributed.

Finally, the residual plot and normal probability plot of the transformed and reduced model (Fig. 31) show that the model assumptions are satisfied. In the residual plot, we see that the variances of the residuals are approximately constant and independent, since there is no (strong) pattern in the plot. The normal probability plot shows that the residuals are approximately normally distributed. There is one potential outlier (shown in red in the residual plot) in this data set. This value corresponds to the oak tree species *Quercus tomentella Engelm* which is the only species in this experiment which grows on an island (Guadalupe). Since this fact explains the unusual behavior of the data value, we may decide not to remove it from the data set.

Interestingly, if the outlier value were removed from the analysis, it would make a drastic difference for the R^2 value (which increases to $R^2 = 0.386$). The p-values for the predictors LN(RANGE) and REGION would become smaller if the outlier value were removed, making a stronger statement about the significance of these factors for ACORN size.

7.2   ANOVA

Analysis of variance (ANOVA) models are used to compare the behavior of a quantitative response for different values of one or more categorical predictors (Section 3.1).

Example 7.4

In microarray experiments, the quantitative response is gene expression or intensity. Suppose an experiment considers the differential expression of genes between two strains of mice. Tissue is collected from three male and three female individuals of each strain. Meaningful questions that can be asked from this experiment are as follows: Are there genes that are differentially expressed between the two strains? Are there genes that are differentially expressed between the two genders? Are there genes for which the expression difference between strains depends on gender? In this experiment, gene expression is the response and there are two categorical predictors (strain and gender), each with two levels. The levels of gender, for example, are male and female.

Depending on the number of categorical predictors, ANOVA models are called one-way (or one factor) for one categorical predictor, two-way (or two factor) for two categorical predictors, etc. In this Manual, we mostly consider one-way and two-way ANOVA models and briefly discuss the four-way ANOVA model commonly used in the analysis of microarray data.

7.2.1   One-Way ANOVA Model

Suppose that a categorical predictor variable takes on I different levels (Fig. 32). Suppose further that the response has mean μ_i when

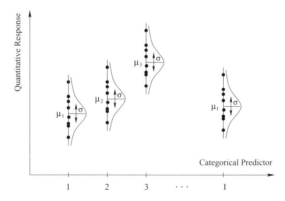

FIGURE 32. In the one-way ANOVA model, observations at predictor level i (dots) are modeled as normal random variables with mean μ_i and standard deviation σ. Notice that the standard deviation is assumed to be the same for all predictor levels.

the experiment is conducted at level i. For each level $i = 1, \ldots, I$ of the predictor variable n_i, observations are collected on a quantitative response. Let x_{ij} denote the jth observation at predictor level i. We assume that the observations on level i are normally distributed with mean μ_i and standard deviation σ. The standard deviation σ is assumed to be the *same* for all predictor levels.

The one-way ANOVA model is:

$$x_{ij} = \mu_i + \epsilon_{ij}, \quad \text{where } \epsilon_{ij} \sim \text{Normal}(0, \sigma^2)$$

The parameters of this model are $\mu_1, \ldots, \mu_I$, and σ.

The ANOVA F-test answers the question of whether or not the response means are the same for all levels of the predictor.

$$H_0 : \mu_1 = \mu_2 = \cdots = \mu_I$$

Biologically, this null hypothesis corresponds to the pessimistic statement that the predictor has no influence on the (average) response.

In Excel: Write the response observations for each predictor level into one column of an Excel spreadsheet. Label the columns by the predictor level if you like. Click TOOLS → DATA ANALYSIS → ANOVA SINGLE FACTOR. Highlight all columns that contain your data (check-mark COLUMNS and LABELS IF APPROPRIATE.)

In the summary table of the Excel output (Fig. 33), you find the group averages and variances. Take a preliminary look at the variances, as they should not be too different in magnitude from each other to satisfy the ANOVA model assumptions (Section 7.2.3). The ANOVA table contains the F-test statistic value and the corresponding p-value for testing whether the response means are all equal. If the p-value is small (less than 0.05), at least one of the predictor levels has a different mean response than the others. The mean standard error (MSE) in the ANOVA table is the estimate of the error variance σ^2.

If an ANOVA F-test concludes that the response means are *not* all equal for different levels of the predictor, it is often of interest to question which means actually differ. This can be done with a Tukey test (Section 6.2.4).

Anova: Single Factor						
SUMMARY						
Groups	*Count*	*Sum*	*Average*	*Variance*		
Column 1	n_1		$\bar{x}_1$	variance within group 1		
Column 2	n_2		x_2	variance within group 2		
Column 3	n_3		$\bar{x}_3$	variance within group 3		
ANOVA						
Source of Variation	*SS*	*df*	*MS*	*F*	*P-value*	*F crit*
Between Groups	SSG	I-1	MSG	F-test statistic	p-value	critical value
Within Groups	SSE	N-1	MSE			
Total	SST	N-1				

FIGURE 33. Schematic of the Excel output for a one-way ANOVA analysis.

Example 7.5

Several studies have linked cigarette smoking in pregnant women to lower infant birth weights. Researchers interviewed 31 women about their smoking behaviors and weighed their newborn babies. The smoking behavior was classified as currently smoking (i.e., smoked during pregnancy), previously smoking (i.e., smoked previously, but quit before pregnancy), and never smoked. The infant birth weights are reported in pounds in Table 3. Performing a SINGLE-FACTOR ANOVA analysis in Excel will produce the output shown in Figure 34.

To interpret the output, look at the average birth weights for the three groups. Even though the average for currently smoking mothers is about 0.4 lb less than for non-smoking mothers, we cannot conclude that there is a statistically significant ($p = 0.502$) difference among the average birth weights based on the data in this study. The reason for this is that the differences are rather subtle and do not

TABLE 3. Baby birthweights (in lbs) of mothers who were smoking at the time of birth, mothers who had smoked previous to pregnancy, and non-smoking mothers.

Current smoker	5.7	7.9	6.8	6.1	7.2	6.2	6.9	6.0	8.4
	7.9								
Previous smoker	7.5	6.8	6.9	5.7	7.9	7.6	8.3		
Non-smoker	7.6	6.9	7.0	6.8	7.8	7.7	6.4	7.4	8.2
	8.6	7.5	7.5	7.5	5.6				

Anova: Single Factor							
SUMMARY							
Groups	Count	Sum	Average	Variance			
Current smoker	10	69.22538	6.92253798	0.8745988			
Previous smoker	7	50.726493	7.24664189	0.73746237			
Nonsmoker	14	102.54987	7.32499094	0.5600545			
ANOVA							
Source of Variation	SS	df	MS	F	P-value	F crit	
Between Groups	0.988070558	2	0.49403528	0.70659848	0.50190463	3.34038556	
Within Groups	19.57687185	28	0.69917399				
Total	20.56494241	30					

FIGURE 34. Output produced by a SINGLE-FACTOR ANOVA analysis in Excel.

show up in a study of this magnitude. Increasing the sample sizes in all three groups would make the study more powerful and would likely lead to significant p-values if there truly is an effect of smoking on birth weight.

To check whether our conclusions are justified, check the assumptions of the one-way ANOVA model. The data should be normally distributed in all three populations. This can be checked via PP-plots (which should be interpreted cautiously, since the sample sizes are not very large). Furthermore, check the variances in the three groups to make sure that they are of comparable magnitude. Here, the largest variance is 0.87, which is no more than twice the smallest (0.56). Thus, we can conclude that the assumptions are satisfied.

7.2.2 Two-Way ANOVA Model

Assume that there are two categorical predictor variables A and B. Variable A takes on I different levels and variable B takes on J different levels. Let μ_{ij} denote the population mean of the quantitative response for factor combination level i of factor A and j of factor B. Let x_{ijk} denote the kth observation on factor combination ij, where $k = 1, \ldots, n_{ij}$. Theoretically, the sample sizes n_{ij} for different factor level combinations can be different; however Excel can only handle the case where the number of observations is the same on every factor level combination.

The two-way ANOVA model is:

$$x_{ijk} = \mu + \alpha_i + \beta_j + \gamma_{ij} + \epsilon_{ijk}, \quad \text{where } \epsilon_{ij} \sim \text{Normal}(0, \sigma^2)$$

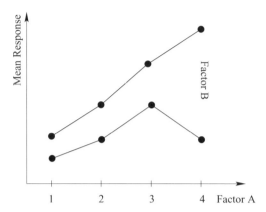

FIGURE 35. In a means plot, the average response values are plotted for all possible factor level combinations. In this example, factor A is plotted on the x-axis with four levels and factor B is plotted as differently colored curves with two levels.

The parameters of this model are the overall mean response μ, the factor effects α_i for factor A, the factor effects β_j for factor B, the interaction effects γ_{ij} of the two factors, and the common error standard deviation σ.

MEANS PLOT: An informative way to gain information on how the mean response changes if the factors are varied in a two-way ANOVA model is a means plot. For each of the possible IJ factor combinations, compute the average response and plot it as in Figure 35. The factor effects α, β, and the interaction effect γ can be seen in a means plot. There is a factor A effect (some $\alpha_i \neq 0$) if the mean response changes for different levels of factor A. In the means plot, this translates to the response curves not being horizontal (flat). There is a factor B effect (some $\beta_j \neq 0$) if the different curves corresponding to the levels of factor B are not on top of each other. The interaction effect ($\gamma_{ij} \neq 0$) can be seen in a means plot whenever the response curves are not parallel. If the curves *are* parallel, then factor B influences the response at every level of factor A in the *same* way and the interaction effect is $\gamma = 0$.

In Excel: Excel can conduct a two-way ANOVA analysis only if the number of replications is the same for each factor combination. Write your data in table form (one factor across the top and the other one on

continued

the left) and put repeated observations for the same factor level below each other.

	Factor A level 1	Factor A level 2
Factor B level 1	x_{111}	x_{211}
	x_{112}	x_{212}
	x_{113}	x_{213}
Factor B level 2	x_{121}	x_{221}
	x_{122}	x_{222}
	x_{123}	x_{223}

Click TOOLS → DATA ANALYSIS → ANOVA: TWO FACTOR WITH REPLICATION. Highlight your observations including all row and column labels. Enter the number of repeated observations per factor level combination into the ROWS PER SAMPLE box.

The output produced by Excel contains averages and variances for the response values at each factor combination. The averages can be used to draw a means plot (recommended) for the ANOVA model. The variances should be of comparable magnitude to satisfy the assumptions of the ANOVA model (Section 7.2.3).

The ANOVA table in the output contains F-test statistics and corresponding p-values that test whether the main effects α and β and the interaction effect γ are equal to zero. A small p-value means that the corresponding effect is *not* equal to zero and is thus important in the model. The MSE (mean squared error) is the estimate of the residual variance σ^2.

Example 7.6

Equal amounts from two strains of yeast *Saccharomyces cerevisiae* and *Saccharomyces exiguus* are grown under three different temperature settings ($25°C$, $35°C$, $45°C$) in a sugar solution growth medium. Growth is measured as CO_2 production (in mL) over a 1-hour period. For each strain of yeast and at each temperature, the experiment is repeated three times. The data are organized in table form (Table 4).

This experiment can be described by a two-way ANOVA model in which growth is the quantitative response, and strain and temperature are two categorical predictors with two and three levels,

TABLE 4. Data for the yeast growth experiment.

	25°	35°	45°
S. cerevisiae	12	37	38
	12	38	42
	11	39	40
S. exiguus	7	19	18
	8	22	16
	8	23	19

respectively. The results of a two-way ANOVA analysis performed in Excel with three replications are:

Anova: Two-Factor with Replication						
SUMMARY	Temp 25	Temp 35	Temp 45	Total		
cerevisiae						
Count	3	3	3	9		
Sum	35	114	120	269		
Average	11.6666667	38	40	29.8888889		
Variance	0.33333333	1	4	188.861111		
exiguus						
Count	3	3	3	9		
Sum	23	64	53	140		
Average	7.66666667	21.33333	17.6666667	15.5555556		
Variance	0.33333333	4.333333	2.33333333	39.2777778		
Total						
Count	6	6	6			
Sum	58	178	173			
Average	9.66666667	29.66667	28.8333333			
Variance	5.06666667	85.46667	152.166667			
ANOVA						
Source of Variation	SS	df	MS	F	P-value	F crit
Sample	924.5	1	924.5	449.756757	7.0277E-11	4.74722534
Columns	1536.11111	2	768.055556	373.648649	1.5582E-11	3.88529383
Interaction	264.333333	2	132.166667	64.2972973	3.8661E-07	3.88529383
Within	24.6666667	12	2.05555556			
Total	2749.61111	17				

To interpret the output, first look at the response variances for all six experimental conditions. They are (approximately) the same magnitude, ranging from 0.33 to 4.33. Since there are only three observations per group, we do not need the variances to be very close. Next, look at the p-values in the ANOVA table. All three values (in rows labeled Sample, Columns, and Interaction) are small (< 0.05). This lets us conclude that the strain (denoted by samples) and the temperature (denoted by columns) both have an influence on the growth rate of yeast. Furthermore, both strains do not react in the same way to changes in temperature, since the interaction effect is also significant. The estimate of the residual variance σ^2 in this experiment is 2.056.

7.2.3 ANOVA Assumptions

The assumptions that are made in every ANOVA model, regardless of how many factors are included in the model, are that the residual terms ϵ are independent and normally distributed with mean zero and constant variance σ^2.

Theoretically, the best way to test these assumptions is to obtain the residuals and then check them for Normality and equal variance, similar to the procedure described in the regression analysis section (Section 7.1.6). Unfortunately, Excel does not provide the residuals for a two-way ANOVA analysis.

An acceptable substitute is to assure that the response measurements are independent (for instance they should not be taken on the same subjects) and to compare the magnitudes of the variances for all IJ possible factor combinations that are computed by Excel. Most of the variances should be similar to each other. However, if the number of replicates per factor level combination is small (≤ 5), it is unavoidable that some will have small variances. Thus, having some variances much smaller than the others is no big reason for concern. The two-way ANOVA model is relatively robust against mild violation of the assumptions.

7.2.4 ANOVA Model for Microarray Data

In microarray experiments, the quantitative response is gene expression as measured by the fluorescence of dyes attached to target cDNA molecules. Categorical predictors commonly used in microarray experiments are

- ARRAY: If more than one slide is used in the experiment, small manufacturing differences can have an influence on the response.

- DYE: In two-color experiments, the dye color (red or green) may have an influence on the magnitude of the response.

- TREATMENT: If two (or more) conditions are compared in the experiment, then it is often the intent to discover the effect that the treatment(s) has on the gene expression response.

- GENE: Naturally, different genes have different levels of expression in any given tissue sample.

Using these four factors, a four-way ANOVA model can be built. The response Y in this model is usually the log-transformed background corrected expression.

$$Y_{ijkgr} = \mu + A_i + D_j + T_k + G_g + AG_{ig} + DG_{jg} + TG_{kg} + \epsilon_{ijkgr}$$

Thus, Y_{ijkgr} is the logarithm of background corrected intensity for the rth repetition of gene g under treatment k labeled with dye j on array i. The interaction effects (array-gene, dye-gene, treatment-gene) reflect the possibility that the gene probes on different slides may differ (AG), that the two dyes have different affinity to bind to certain sequences (DG), and that genes may be expressed differently under different treatments (TG). Note that other interaction effects are missing from the model. For instance, there is no array-dye (AD) interaction effect, as the dyes are assumed to function the same on each slide.

In the context of the above ANOVA model, the null and alternative hypotheses corresponding to differential expression of gene g under treatments 1 and 2 become

$$H_0 : T_1 + TG_{1g} = T_2 + TG_{2g}$$
$$H_a : T_1 + TG_{1g} \neq T_2 + TG_{2g}$$

Statistical software (unfortunately not Excel) will provide a p-value for this hypothesis test. If there is a large number of genes in a microarray experiment, many p-values will need to be considered. Since the number of tests in microarray experiments is often very large (in the thousands), additional considerations need to be made to control the number of false-positive decisions (Section 8.4.6).

7.3 What ANOVA and Regression Models Have in Common

The distinction that is made between ANOVA and regression models is somewhat artificial. Technically, these two statistical models are equivalent. They both model a quantitative response as a linear function of one or more predictor variables. They both assume that the residuals are independent, normally distributed, and have constant variance. The only difference between them is the nature of the predictor variables. However, it is possible to include categorical variables in regression models by coding the different levels through DUMMY variables. This becomes especially easy if the categorical predictor variable takes on only two levels (code them 0 and 1, for example).

Example 7.7

If a categorical predictor takes on more than two levels, more care must be taken before it can be included in a regression model. For instance, if a drug is administered at doses 2, 4, and 8, then the dummy variable representing drug dosage in the regression model *should not* be coded 2, 4, and 8 (if it cannot be assumed that the dosage has a linear influence on the response). Likewise, the drug dosage should also not be coded 0, 1, and 2 because this would assume that the change in response is the same if switching the dosage from 2 to 4 as it is when switching the dosage from 4 to 8. Instead, use two dummy variables, both coded 0 and 1, to represent the drug dosage.

dosage represented	dummy variable x_1	dummy variable x_2
2	0	0
4	1	0
8	1	1

In this model, the slope associated with the dummy variable x_1 describes the change in response corresponding to a change in dosage from 2 to 4, and the slope associated with the dummy variable x_2 describes the change in response when changing from 4 to 8. Consequently, the sum of the two slopes describes the change in response when changing the dose from 2 to 8.

8   Special Topics

In this final chapter, we describe special topics in statistics most commonly used in biology. The first two popular topics are classification and clustering. Both techniques are used, for example, in evolutionary biology to create phylogenies, to specify co-expressed genes in molecular biology, or to find functionally related proteins (Sections 8.1 and 8.2). Another popular technique used in biology that we describe is principal component analysis (PCA). PCA is a statistical procedure that makes high-dimensional data (with observations collected on many variables) easier to interpret by reducing it to fewer dimensions while conserving the most important aspects of the data (Section 8.3). This procedure is routinely applied in the analysis of microarray data, where it is common to observe thousands of variables (genes). Microarray data analysis itself is a special topic in statistics that is described. The high-dimensional nature of microarray data gives rise to many new statistical challenges; among those described here is the multiple comparison problem, i.e., the task of statistically addressing many questions from the same data set (Section 8.4). Another special topic that is addressed is maximum likelihood estimation. This statistical technique is commonly used to estimate the values of model parameters from observations or data (Section 8.5), and has benefited greatly from the increased computational capacity that is now widely available. This technique is one of the most common approaches that is used when dealing with complex biological models (for instance in computational phylogenetics or bioinformatics). Finally, we contrast the frequentist procedures described in this book with alternative Bayesian methods and highlight important differences between the two approaches (Section 8.6).

8.1   Classification

Many large-scale studies on humans have been conducted to better understand what conditions cause a particular disease, such as cancer or heart disease. Data are collected from many patients in the form of genetic information (e.g., microarrays), physiological measurements, and surveys about lifestyle, family history, etc. A typical goal is to predict the chance that an individual with a particular background (genetic and lifestyle) will develop the disease under study. The task of separating a set of n-dimensional observations into different groups (e.g., high risk, medium risk, low risk) or allocating a new observation into previously defined groups is one of STATISTICAL CLASSIFICATION.

Example 8.1

Indian elephants have smaller ears (on average) than their African cousins. However, ear size by itself (measured in diameter or area) is not a good classification tool to determine whether an elephant is of Indian or African origin. If ear size were used as the single determining factor, then all baby elephants would (naively) be classified as Indian, and only when they grew up would some of them be classified as African. Instead, ear size in conjunction with some other variable (such as age or height) would be a much better tool for classifying the animals.

There are two main goals in statistical classification.

- To describe either graphically (in three or fewer dimensions) or algebraically (in higher dimensions) the differential features of observations from several known populations. The goal is to find a DISCRIMINANT FUNCTION of the observations that describes the differences well (Fig. 36a). This task is usually referred to as statistical DISCRIMINATION.

- To sort observations into two or more labeled classes. The objective here is to develop a CLASSIFICATION RULE that can be used to sort observations and that has minimal misclassification error (Fig. 36b). This is typically referred to as statistical CLASSIFICATION.

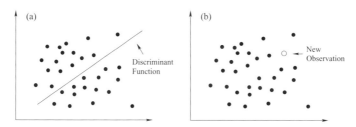

FIGURE 36. The goal of classification is to find a discriminant function that divides existing data by type while minimizing the percentage of misclassified observations. New observations can then be allocated to a particular type using this discriminant function.

All classification problems are based on sets of TRAINING DATA. Those are n-dimensional observations whose types are known. These may include measurements on patients with and without a particular disease, ear diameter and height of elephants of Indian and African origin, or measurements on other populations which are categorized by a predictor whose value is observed. Based on this set of training data, a classification rule is developed (often in the form of a discriminant function) according to which new observations may then be classified into one of the existing categories.

Figure 37a shows a linear discriminant function that adequately separates two types of two-dimensional observations. Higher-dimensional observations ($n > 2$) on two different types may also be separated by a linear discriminant function (a plane in three dimensions or a hyperplane in higher dimensions). However, in some applications, non-linear discriminant functions may be more appropriate than linear ones (Fig. 37b). As is the case with almost any statistical model, the best discriminant function is the result of a trade-off between simplicity of the model (linear is simpler than higher-degree polynomials) and low classification error (percentage of misclassified data points from the training).

8.2   Clustering

In statistical classification problems, training data are available for which the classification category of the observations is known. In some biological applications, this is not possible. For example, for gene expression experiments on the yeast cell cycle, the exact function of a majority of genes is not known. Instead, expression patterns

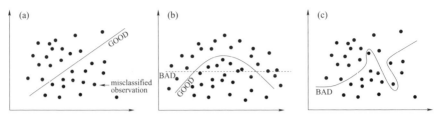

FIGURE 37. The goal of classification is to minimize classification error using the simplest possible discriminant function. Here, the data consist of two-dimensional observations of two types (red and black). (a) A linear discriminant function is appropriate, even though two observations from the training set are misclassified. (b) A curvilinear discriminant function separates the data very well and is preferable to the linear function which has too many misclassified observations (dotted line). Even though the curvilinear function in (c) has no classification error on the training set, it is much too complicated to be preferable to the simpler version from a.

are observed repeatedly across many organisms such that groups of genes can be formed which exhibit similar behavior.

Sorting n-dimensional observations into k groups such that the members of each group are similar to each other and dissimilar from members of other groups is the goal in statistical CLUSTERING (Fig. 38). One challenge is that the number k of groups is not necessarily known before the experiment is conducted. What "similar" means in a particular experiment depends on the measurements that were

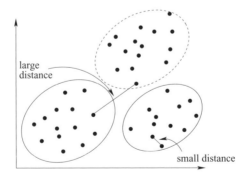

FIGURE 38. Clustering is the statistical task of separating n-dimensional observations into meaningful subsets. Here, two-dimensional observations of unknown type are separated into three clusters based on their proximity (Euclidean distance). Distances between observations in the same cluster are small compared to distances between observations in different clusters.

taken. A DISTANCE MEASURE is needed that can describe how dissimilar two observations are.

Example 8.2

For n-dimensional quantitative observations, such as gene expression on n observed genes, a possible distance measure is Euclidean distance in n-dimensional space. Suppose that $\vec{x} = (x_1, x_2, \ldots, x_n)$ are data observed from n genes collected on individual A, and $\vec{y} = (y_1, y_2, \ldots, y_n)$ are n observations on the same genes collected from individual B. Then the Euclidean distance for these two vectors of observations is

$$d(\vec{x}, \vec{y}) = \sqrt{(x_1 - y_1)^2 + \cdots + (x_n - y_n)^2}.$$

If there is prior knowledge about gene expression in the organism of study, then another sensible distance measure may weigh gene expression differences according to their variability. An observed difference in genes which typically does not vary much should count as "more different" than the same observed difference in a highly variable gene. The corresponding statistical distance measure is called the MAHALANOBIS DISTANCE.

There are many other distance measures that may be used to describe similarities and dissimilarities between multi-dimensional observations. They can be based on absolute values of coordinate differences, maximum coordinate distances, or similar measures. If some of the observations are categorical, other creative measures may need to be employed. All distance measures have two features in common. First, they are equal to zero if the observations on both individuals are identical. Second, larger distance measures reflect observations which are less similar to each other.

There are two fundamentally different approaches to the task of statistical clustering of n-dimensional data:

- HIERARCHICAL CLUSTERING: Starting with either single observations (AGGLOMERATIVE CLUSTERING) or with one big set containing all observations (DIVISIVE CLUSTERING), the distance measure is used to form or break up groups of observations. In agglomerative clustering, initially, every observation is its own cluster. Similar observations (those with the smallest distance measure) are clustered together (Fig. 39).

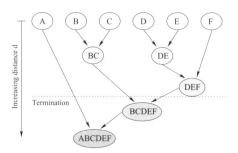

FIGURE 39. In agglomerative hierarchical clustering, observations are combined according to degree of similarity (most similar observations get combined first). In the above example, the similarity between B and C and between D and E is the same. The DE cluster is most similar to F etc. The algorithm may be terminated at any time, yielding the currently existing clusters (A), (BC), and (DEF) as a result.

The results from clustering can be displayed in the form of a tree diagram. Depending on when the algorithm is terminated, the results will be either many small clusters (terminate early) or fewer larger clusters (terminate later).

- PARTITIONAL CLUSTERING: In this method, the number, k, of clusters remains fixed and observations are "shuffled" between clusters until an optimal clustering configuration has been reached. One challenge of this approach is to determine the optimal number of clusters that should be used.

8.2.1 Hierarchical Clustering

The main decision that needs to be made in hierarchical clustering is the choice of which two clusters to "fuse" (or split) at each stage of the developing tree diagram. Since the processes behind agglomerative clustering and divisive clustering are identical apart from the direction (top-down or bottom-up) in which the tree is developed, we concentrate only on agglomerative clustering here.

Suppose the data consist of observations on N individuals. Each observation is recorded in the form of an n-dimensional vector. At the first stage of the process, each individual is its own cluster. Pairwise dissimilarity measures d are computed for all possible pairs of individuals and recorded in the form of an $N \times N$ table. The two individuals with the smallest distance are merged together into a cluster.

In the next step, we start with $N - 1$ clusters. Again, we need distances between all possible pairs of clusters to fuse the most similar clusters together. How should similarity measures for groups of observations be computed? There are three different alternatives.

To compare two clusters, we need a measure that compares the elements that are already in the clusters. Three different methods are commonly used in practice:

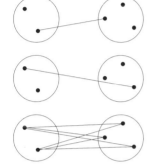

- SINGLE LINKAGE: Minimum distance, or distance between nearest neighbors.

- COMPLETE LINKAGE: Maximum distance, or distance between farthest neighbors.

- AVERAGE LINKAGE: Average distance, or average of all pairwise distances.

Note: All hierarchical clustering procedures follow essentially the same protocol: Select a distance measure and a linkage type.

1. Compute the table of pair-wise distances.

2. Fuse the two clusters with highest degree of similarity (smallest distance) and reduce the number of clusters by 1.

3. Record the identity of all clusters and the levels (distances) at which mergers take place.

4. Recompute distance table in smaller dimension using appropriate linkage and repeat.

The result of a hierarchical clustering can be drawn in a tree diagram with subject labels on one axis and the distance measure d on the other axis.

Example 8.3

Nei and Roychoadhury (1993) studied evolutionary distance between humans of different racial origins. They studied 26 human populations and described the genetic distance between pairs of populations via Cavalli-Sforza distance. For a small subset of populations under study, the genetic distances (multiplied by 100) are:

	Ge	It	Ja	Ko	Ir
German		0.6	5.7	5.7	1.8
Italian	0.6		5.5	5.5	5.0
Japanese	5.7	5.5		0.6	5.0
Korean	5.7	5.5	0.6		5.2
Iranian	1.8	1.6	5.0	5.2	

To create a phylogenetic tree of these populations using single linkage, notice that the most closely related populations are German and Italian (distance 0.6) as well as Japanese and Korean (distance also 0.6). Joining these two pairs of populations yields three new clusters. Next, the new distances between clusters (using single linkage) need to be recomputed.

	Ge/It	Ja/Ko	Ir
German-Italian		5.5	1.6
Japanese-Korean	5.5		5.0
Iranian	1.6	5.0	

For example, the new distance between the German-Italian and Japanese-Korean clusters is 5.5, because this is the shortest distance between any two members of the two clusters. Next, the Iranian cluster is joined with the German-Italian cluster, because it has the smallest dissimilarity measure ($d = 1.6$). Finally (not shown in table), the single linkage between the German-Italian-Iranian and Japanese-Korean clusters is $d = 5.0$. The resulting tree diagram is shown in Figure 40. The ordering of the populations on the y-axis of the diagram is arbitrary. Typically, populations are ordered so that the branches of the tree do not overlap.

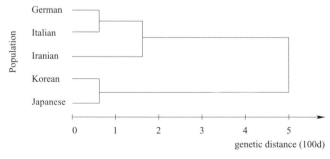

FIGURE 40. A tree diagram illustrating the genetic differences between five selected human populations. The tree is based on the Cavalli-Sforza dissimilarity measure d and single linkage.

8.2.2   Partitional Clustering

One of the most popular methods of partitional clustering for quantitative data is called K-MEANS CLUSTERING, in which the complete data set is partitioned into k clusters by shuffling the observations around in a way that will improve the separation between clusters until a termination criterion is reached.

ALGORITHM:

1. Choose a number of clusters k.
2. Randomly split the data into k (non-empty) clusters. Determine the center of each cluster by averaging the observations in the cluster.
3. Assign each data point to its nearest cluster center.
4. Compute new cluster centers and repeat.
5. Terminate after convergence (usually when the assignment of individuals to clusters does not change).

This algorithm is simple to implement, runs reasonably fast even for very large data sets, and does not require storage of intermediate results. Since the results depend on the number k of clusters specified in the algorithm and the initial configuration of observations, it is a good idea to repeat the algorithm with different k and a different initialization to determine the robustness of the results.

> **Note:** Obtaining a reasonable estimate for the number, k, of clusters that should be created is a challenging problem that is beyond the scope of this Manual. Note that the k-means clustering algorithm will always return results regardless of whether the specified k is biologically meaningful or not. However, the resulting clusters and subsequently their interpretation in a biological context, may change dramatically with different choices of k.

Example 8.4

Consider the four two-dimensional observations shown in Figure 41:

Item	Observations x_1	x_2
A	1	2
B	5	2
C	2	3
D	7	1

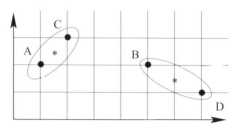

FIGURE 41. Four two-dimensional observations are clustered using the
k-means clustering algorithm. Observations are depicted by red dots and
final cluster centers are depicted by stars.

k means clustering (with $k = 2$) is used to cluster the observations.
Initialize the algorithm by randomly forming two groups of observations. For instance, consider the initial clusters (AB) and (CD).
The numbers of observations in the initial clusters do not need to be
equal, but every cluster must contain at least one observation. Next,
compute the centers of each cluster by averaging coordinates:

$$AB \text{ center: } \left(\frac{1+5}{2}, \frac{2+2}{2}\right) = (3, 2)$$

$$CD \text{ center: } \left(\frac{2+7}{2}, \frac{3+1}{2}\right) = (4.5, 2)$$

Now consider the observations, one by one, and find their nearest
cluster center. Observation A is closest to the (AB) cluster center,
so no action is taken. However, observation B is closer to the (CD)
cluster center, rather than the (AB) cluster center. Therefore, we
shift this observation to form new clusters. The cluster centers now
have to be recomputed:

$$A \text{ center: } = (1, 2)$$

$$BCD \text{ center: } \left(\frac{5+2+7}{3}, \frac{2+3+1}{3}\right) = (4.67, 2)$$

Check distances of observations from centers again. A, B, and D are
closest to the cluster that they are in. But observation C is now
closer to the A-cluster center. Therefore we have to shift observation
C and recompute centers one more time. Now, every observation is
closest to its own cluster center and no more shifts are necessary.
The algorithm is terminated and yields the final clusters (AC) and
(BD).

8.3   Principal Component Analysis

When we think about variables (or multiple dimensions), it is very easy to consider a single variable or dimension (e.g., time) by itself. Thinking about two variables at the same time, say, plant development across time, is not that difficult, nor is it difficult to represent graphically (i.e., $x - y$ plot). Even though the correlation (or covariation) between two predictor variables is not captured in a two-dimensional plot, it is simple to calculate, given the tools that have been described in Section 3.8. However, when the number of variables (e.g., all genes in a genome) in an experiment becomes large, it is extremely difficult to think about all the relationships among the variables and how they are correlated. Luckily, there are a number of exploratory statistical techniques that can be used that both reduce the dimensionality of the data and allow the visualization of trends. In other words, instead of thinking about, say 5 variables at the same time, it is possible to think about these 5 predictor variables, or 5 dimensions, by forming a linear combination of 5 predictor variables via a statistical model. Together in a linear model, the subset of predictor variables explains the largest amount of variation in a response variable.

Principal component analysis (PCA) is a statistical technique that surveys the dimensionality of the data space, as defined by the collection of observed variables centered around their means. It takes advantage of the variation in and between variables (i.e., correlation) to transform the original (correlated) data to an informative set of uncorrelated variables that together explain the largest portion of the variation in the response variable. As it turns out, PCA operates from the covariance matrix by calculating the eigenvectors. Without going into an explanation of linear algebra (i.e., projections, eigenvalues, eigenvectors, etc.), PCA is known to be the optimal linear transformation of the data space and is able to produce the subspace (i.e., representation) that has the largest variance between components. A linear transformation is presented as a function of the variables that are each weighted by what is referred to as a "loading." The first principal component explains the most variability in the data through its loadings on the variables in the model. The second principal component explains less variation in the data space, while the third and remaining principal components explain less and less of the variation. Typically in dimension reduction applications, the majority of the variation in the data are explained by the first few principal components. PCA is known as a multivariate technique in the field of statistics, since it operates on multiple dimensions (vectors) in the data space. It is a standard statistical analysis that is available in most software packages.

8.4   Microarray Data Analysis

From a statistician's perspective, microarray data or data generated by other high-throughput sequencing technologies provide a number of new challenges that do not occur in more conventional biological data sets (Craig et al. 2003). One challenge is that the number of treatment combinations (different experimental conditions applied to a large number of genes on physically different arrays, possibly treated with different dyes, etc.) is typically very large (easily in the thousands, often in the tens of thousands). In contrast, the number of identical replications in which the experimental conditions stay exactly the same and are applied to the same biological sample (technical replicates), or the number of replications in which the experimental conditions stay the same and are applied to different biological material (biological replicates), are usually very small. Sometimes there may be as few as two or three biological and/or technical replicates compared to tens of thousands of experimental conditions. This makes statistical analysis challenging, because it relies on comparing variation across replications to differences between experimental conditions (Kerr and Churchill 2001).

8.4.1   The Data

Most biologists receive the data from their microarray experiments in spreadsheet format. The data files contain details about the information spotted on the arrays and their physical positions, as well as per-spot laser (light) intensities that represent the amount of corresponding mRNA measured in the sample. Since microarray technologies vary, the array platform used in the experiment dictates the format of the data and the way that mRNA quantities are represented.

OLIGONUCLEOTIDE ARRAYS

These commercially available arrays are the most common microarray platform. Probes on arrays are short oligonucleotide chains (e.g., 25-mer on Affymetrix arrays and 60-mer on Agilent arrays) that each correspond to *a portion* of a gene. Several (between 11 and 20) of these probes together identify a gene in a probe set. To identify the degree of cross-hybridization, the "perfect match" probes on Affymetrix arrays are accompanied by "mismatch" probes in which the middle of the oligonucleotide is deliberately altered. The "mismatch" measurement is subtracted from the "perfect match" measurement for every probe. To obtain an expression measurement for each gene, the mismatch-corrected intensities for all probes corresponding to a gene may be averaged.

SPOTTED ARRAYS

Spotted arrays are typically customized arrays that are used in smaller research applications. Every spot on the array is a sequence that corresponds to a specific mRNA. Typically, two biological samples are mixed together and then hybridized to a single array. To distinguish between the samples, the biological material is labeled with different fluorescent dyes (Cy3, green and Cy5, red). A laser is used twice to measure fluorescence, once for each color using different wavelengths of light. Each laser scan provides data from a variety of pixels that represent each spot on the array. The data file usually contains the mean and median intensities for the center (foreground) pixels as well as the perimeter (background) pixels for both colors. Typically, the file also contains columns representing the "background corrected" median intensities, which are the background medians subtracted from the foreground medians for each spot.

In almost all microarray experiments, two biological samples are compared to each other. The two samples may be hybridized to the same spotted array or to two different oligonucleotide arrays. For simplicity, we denote the measurements for gene g (where g ranges over the possibly thousands of probes or genes represented on an array) by R_g and G_g, where R_g and G_g are the red and green fluorescence, respectively. On a one-color oligonucleotide chip, R_g and G_g should be understood to be the fluorescence of a probe g for two conditions, each on two different chips.

To compare the measurements for two samples across thousands of genes, it is convenient to first transform the measurements into the following format:

$$M_g = \log_2 \frac{R_g}{G_g}$$

$$A_g = \frac{1}{2} \left(\log_2 R_g + \log_2 G_g \right).$$

M represents the log-fold change of background corrected red intensity compared to green intensity for gene g. The $\log_2$ assures that the M-value is equal to 1 if the red intensity is twice as large as the green. On the other hand, if the green intensity is twice as large as the red, then the M-value is -1. If there is no difference in expression between the red and green conditions, then the M-value is zero. Where the M-value represents the relative difference in expression between the two samples, the A-value expresses the average log-intensity of the two measurements. If gene g is not (or only slightly) expressed under both experimental conditions, then A will be small. If the gene is highly expressed under at least one condition, then A will be large.

Before any kind of analysis is conducted, most researchers plot the M-values for all genes on an array against the respective A-values in an MA-plot to gain a visual overview of the data (Fig. 42a).

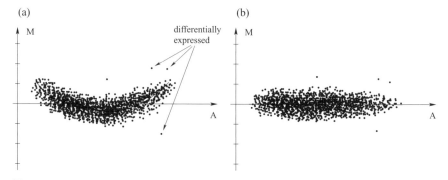

FIGURE 42. MA-plots help to visualize the results of a microarray experiment. The genes which are most likely differentially expressed are those with both large M-values and large A-values. Large M-values combined with small A-values may be explained through technical variation rather than actual biological differences. An MA-plot may reveal that there is a dependence of the fold-change (represented by the M-value) on the intensity (represented by the A-value) (a). A LOESS normalization may be employed (b) to resolve this problem.

8.4.2 Normalization

Before any conclusions can be drawn from microarray data, it is essential to mathematically remove as much systematic variation *not* caused by biological effects as possible. This process is called Normalization. Possible sources of systematic technical errors in the data may be caused by manufacturing differences in the arrays themselves, the dyes used to label the samples, the print tips with which the slides were spotted, etc. Each of these sources may be addressed with different types of normalization methods. Methods may be combined and more than one normalization technique can be applied to the same data (Yang et al. 2002). The most common normalization techniques are discussed below.

GLOBAL NORMALIZATION: Consider a whole-genome gene expression study that compares every gene in the genome under a treatment condition and under a control condition. If it makes biological sense to assume that there should be approximately as many up-regulated genes as down-regulated genes, then this assumption is equivalent to the mathematical assumption that the average of all M-values is zero. In an MA-plot this means that the point cloud should be centered vertically at zero. In real microarray data this is sometimes not the case. If, for instance, the expression values of one condition are systematically inflated (due to a difference in the amount of material hybridized, a different laser setting, a problem with the dye, etc.), it will lead to the average of the M-values being non-zero. A global normalization computes the average of all M-values from the entire array and subtracts this value from each M-value to force the mean to be zero. In an MA-plot this corresponds to moving the point cloud up, or down to center it at zero, *without* changing the shape of the point cloud.

LOESS NORMALIZATION: LOESS (or LOWESS) stands for locally weighted scatterplot smoothing. In microarray experiments it is often the case that the MA-plot exhibits some sort of shape that is *not* a random point cloud centered around the $M = 0$ axis. For example, in Figure 42a the point cloud has a U-shape. This phenomenon occurs typically on two-color arrays if the dye bias is dependent on the fluorescent intensity. Since such a dependence does not make sense biologically, it is removed by smoothing the point cloud (via adjusting the M-values) so that the resulting point cloud is centered around the $M = 0$ axis (Fig. 42b).

DYE-SWAP NORMALIZATION: In two-color arrays, the two dyes that are often used in microarray experiments (Cy3 and Cy5) have different binding affinities to certain sequences. This makes interpreting results difficult, because it can be unclear whether low fluorescence is caused by low amounts of RNA in the sample or by poor binding affinity of one of the dyes. To correct this problem, dye-swap experiments are often carried out (see Example 8.5). The two samples whose expression values are to be compared are each split in half. One-half of the treatment sample is labeled with red dye and the control with green, and for the other sample halves the dye labeling is reversed (or swapped). Then, green treatment and red control are hybridized on one array and red treatment and green control are hybridized to a second array. M-values are computed for both hybridizations (keeping the order of treatment and control the same). The M-values for each gene may be averaged across the two arrays to remove the dye bias.

PRINT TIP NORMALIZATION: During the manufacturing process of spotted cDNA arrays, genetic material is collected in well plates and robotically printed (or spotted) onto the array surface with fine needles. Multiple needles or print tips are arranged in a grid-like format to simultaneously pick up material and print it on the slide. In the printing process, the print tips may become slightly deformed, resulting in systematic differences in the shape of the spots printed by damaged print tips. The ramification of non-uniform spotting during manufacturing of the arrays results in expression values for the genes printed by the damaged print tip having altered intensity values (i.e., all slightly higher or lower) than those printed by undamaged print tips. To identify differences between tips and to correct systematic non-biological differences, one can create a plot of the M-values for each separate print tip (Fig. 43). If one curve looks very different from the others, then it stands to reason that this effect is due to a damaged print tip, rather than to any true biological differences. Normalization in this case is used to find the average print tip distribution and to adjust M-values so that all have the same print tip curve pattern.

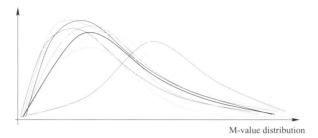

FIGURE 43. Print tip distribution plot. The distributions of M-values are plotted as differently colored curves, one for each print tip used to spot the array. In this case, the print tip represented by the red curve exhibits systematic differences from the other print tips. The curves may be averaged (thicker black line) for normalization purposes.

After normalization, genes can be ranked by the magnitude of their log-fold changes (absolute M-values), and those with the largest log-fold changes are often declared differentially expressed. However, this process of normalizing and ranking genes is somewhat subjective both in the choice of normalization methods to be applied and in the threshold chosen for normalized log-fold changes. Statistical methods, such as an ANOVA model (described in the following section), can replace the normalization and ranking process. If replication has been employed, and per-gene information is available, statistical methods are useful for testing statistical hypotheses, such as differential expression, with greater accuracy.

8.4.3   Statistical Analysis

To make sense of the many observations and attribute "effects" to the treatments that occur in a microarray experiment, an ANOVA model (introduced in Chapter 7) can be used. In this model, every systematic change in the experiment, such as treatment group (one or more different treatment[s] and control), different dyes, each different gene on the array, each physical array used, etc. is enumerated. After enumerating conditions, every observation is given a label. In microarray experiments, the observations can be absolute expressions for a one-color chip or the M-values of ratios of expression for a two-color chip. The technical effects of the experimental conditions and the biological effects of different samples are estimated and compared. Statistical hypotheses, for instance, the hypothesis of differential expression of a gene between two conditions, may be tested using the model (Newton et al. 2001).

Example 8.5

In every experiment, it is important to keep track of conditions that may be varied by the experimenter and to carefully organize the resulting data. This principle becomes especially important in microarray experiments since there are so many observations (possibly under many different treatment combinations) recorded. Consider, for instance, a very simple dye-swap experiment, in which only two genes are observed. The genes are spotted in triplicate on two arrays. In this case, the conditions that we need to keep track of and enumerate are:

- Array: taking on values 1 (for array 1) and 2 (for array 2).
- Treatment: taking on values 1 (for treatment) and 2 (for control).
- Dye: taking on values 1 (for red) and 2 (for green).
- Gene: taking on values 1 (for gene 1) and 2 (for gene 2).
- Replication: taking on values 1, 2, and 3, respectively.

Each observation can be identified or indexed by a combination of the above conditions. For most microarray experiments, the observations themselves are the background corrected median intensities of fluorescence for each spot on each array. The enumeration process is depicted in Figure 44. In reality, this process is of course not carried out by hand but by a statistical software program. Excel is not equipped to handle higher-order ANOVA models such as the ones used for analysis of microarray data.

	A	B	C	D	E	F	G
1	Array 1				Array 2		
2		backgound corrected log-intensity				backgound corrected log-intensity	
3	Gene Name	treatment (red)	control (green)		Gene Name	treatment (green)	control (red)
4	Gene 1	-2	0		Gene 1	1	2
5	Gene 1	-1	1		Gene 1	0	0
6	Gene 1	0	1		Gene 1	-1	-1
7	Gene 2	3	5		Gene 2	10	2
8	Gene 2	4	3		Gene 2	3	1
9	Gene 2	1	7		Gene 2	5	4
10							

FIGURE 44. Every observation in a microarray experiment is labeled with the treatment combinations under which it was derived. For instance, the log-corrected intensity value 3 (circled in red) is the observation taken from array 1 under treatment 1 labeled with dye 1 on gene 2, repetition 1. Every other observed value is labeled in a similar manner.

8.4.4 The ANOVA Model

The statistical model used for the analysis of microarray data seeks to explain as many as possible of the "effects" caused by the different experimental conditions. The enumerated observations Y_{ijkgr} are written as a sum of effects plus a random error term.

$$Y_{ijkgr} = \mu + A_i + T_j + D_k + G_g + AG_{ig} + DG_{jg} + TG_{kg} + \epsilon_{ijkgr}$$

Here, μ is the overall expression average observed in the experiment. A stands for the array effect, T stands for treatment effect, D stands for dye effect, and G stands for gene effect. If, for example, the treatment on average had *no* effect on any of the observed genes, then the treatment effect terms would both be zero ($T_1 = T_2 = 0$). The terms AG, DG, and TG are interaction effects. They express the possibility that a particular treatment may have an effect on some genes, but not on others. Notice that some potential interaction terms (such as the array-dye interaction, AD) are missing from the model. This is because the dyes are assumed to function the *same way* regardless of which array is used in the experiment. The ϵ are error terms that cannot be explained through any systematic change in experimental conditions. They represent the biological variation from individual to individual as well as technical variation through measurement error. The ANOVA model assumes that the errors ϵ are independent and have a normal distribution.

The next step in the statistical analysis of a microarray experiment is to estimate all systematic treatment and interaction effects. This is done by simply averaging the observations. Statisticians use "dot" notation to represent taking averages. For example, averaging the replicates (r) of gene i's expression is represented by

$$Y_{ijkg\bullet} = \frac{1}{\# \text{ of replications}} \sum_{\text{all } r} Y_{ijkgr}$$

The model parameters are all computed by averaging the appropriate set of observations (Table 5).

Example 8.6

In the simple dye-swap experiment described in Example 8.5 there are two treatments, and subsequently, there are two treatment terms

TABLE 5. Model parameters in an ANOVA model
are obtained by averaging observations.

Parameter	Estimate
μ	$\bar{Y}_{\bullet\bullet\bullet\bullet\bullet}$
A_i	$\bar{Y}_{i\bullet\bullet\bullet\bullet} - \bar{Y}_{\bullet\bullet\bullet\bullet\bullet}$
D_j	$\bar{Y}_{\bullet j\bullet\bullet\bullet} - \bar{Y}_{\bullet\bullet\bullet\bullet\bullet}$
T_k	$\bar{Y}_{\bullet\bullet k\bullet\bullet} - \bar{Y}_{\bullet\bullet\bullet\bullet\bullet}$
G_g	$\bar{Y}_{\bullet\bullet\bullet g\bullet} - \bar{Y}_{\bullet\bullet\bullet\bullet\bullet}$
AG_{ig}	$\bar{Y}_{i\bullet\bullet g\bullet} - \bar{Y}_{i\bullet\bullet\bullet\bullet} - \bar{Y}_{\bullet\bullet\bullet g\bullet} + \bar{Y}_{\bullet\bullet\bullet\bullet\bullet}$
DG_{jg}	$\bar{Y}_{\bullet j\bullet g\bullet} - \bar{Y}_{\bullet j\bullet\bullet\bullet} - \bar{Y}_{\bullet\bullet\bullet g\bullet} + \bar{Y}_{\bullet\bullet\bullet\bullet\bullet}$
TG_{kg}	$\bar{Y}_{\bullet\bullet k g\bullet} - \bar{Y}_{\bullet\bullet k\bullet\bullet} - \bar{Y}_{\bullet\bullet\bullet g\bullet} + \bar{Y}_{\bullet\bullet\bullet\bullet\bullet}$

T_1 and T_2 that can be estimated. The overall mean μ is computed
as the mean of all 24 observations (estimate of μ: $\bar{Y}_{\bullet\bullet\bullet\bullet\bullet} = 2$). The
treatment effect is computed by taking the averages of the treatment
(1.9167) and control (2.0833) observations separately and subtracting the overall mean from both. Hence,

$$T_1 = 1.9167 - 2 = -0.0833, \qquad T_2 = 2.0833 - 2 = 0.0833.$$

A statistical hypothesis test can be employed to decide whether these
treatment effects can be explained by random error or whether they
are due to some biological effect.

A hypothesis test for differential expression may use the null hypothesis

$$H_0 : T_1 + TG_{1g} = T_2 + TG_{2g}$$

and alternative hypothesis

$$H_a : T_1 + TG_{1g} \neq T_2 + TG_{2g}.$$

Notice that for real microarray data, a large number of these tests
need to be conducted—one test for every gene g represented on the
array.

8.4.5   Variance Assumptions

Most statistical tests, among them the two-sample t-test (Section
6.2.1) and the ANOVA F-test (Section 6.2.3), used in the analysis
of microarray data rely on comparing the difference in observations
between treatment and control to a variation measure. There are

different methods by which to estimate this variation from the gene expression observations. If each gene is spotted more than once on the array, then it is possible to compute a statistical "per-gene" variance for each array. Most often, however, the number of times that genes are spotted on an array is small (usually less than five). Thus, this variation measure will lack statistical accuracy.

It is possible to "borrow" variation information from other genes. Whether or not this makes biological sense has to be decided on a case-by-case basis. A very broad approach (that most likely does *not* make biological sense) is to estimate the gene expression variance based on all gene observations on the array. This single estimate of variance is referred to as the "common variance" and is often used to individually test every gene for differential expression. The approach of using a common variance assumption is biologically suspect, since regardless of what the treatment actually is, some genes will naturally be highly expressed, while others will show almost no expression. The genes with very low expression will likely have very small variation associated with them, while the expression of highly expressed genes may vary greatly across biological replications or even over technical replications of the same gene on the same array. Using the same variation measure for all genes will thus inflate the test statistic (and with it the p-value) for differential expression for some genes but deflate the value for other genes. A compromise combines variance information only across genes whose expression values are comparable.

8.4.6   Multiple Testing Issues

In all microarray experiments, and in most other experiments in molecular biology, a large number of questions may be posed and answered by the same experiment. For instance, the same question is asked of potentially tens of thousands of genes from the same microarray experiment, namely, "Is *this* gene differentially expressed between treatment and control?" Statistical tests are not infallible. In fact, the significance level α that we use in a statistical test (usually $\alpha = 0.05$) represents our willingness to declare an effect where in reality there is none (see Section 6.1.2). For every test that is conducted, it is possible that the conclusion drawn is wrong. If thousands upon thousands of tests are conducted, then likely there will be many wrong conclusions. The problem is that we do not know *which* of the many conclusions are faulty.

Example 8.7

Consider a microarray experiment in which the expression values of 1000 genes are observed for a treatment condition and a control condition. Suppose there is sufficient replication so that we can conduct 1000 two-sample t-tests to test each gene for differential expression. Even though we will likely be correct in declaring many genes as differentially expressed, if we use a significance level of $\alpha = 0.05$ for each test, then we expect to incorrectly declare 50 genes (5% of 1000) as differentially expressed. The problem is that we do not know which genes are correctly declared as differentially expressed and which genes are "false positives."

How can this problem be addressed? Of course, we could drastically lower the significance level of every test. For instance, if the experimenter in Example 8.7 is willing to go on one wild goose chase, she could work with an α value of 0.001 (0.1% of 1000 is 1). However, this would lead to her declaring very few genes as differentially expressed. In fact, it would make her list of differentially expressed genes much shorter, because it would remove both the "fake" genes as well as some truly differentially expressed genes.

An alternative approach has been suggested by Hochberg and Benjamini (1995). Instead of considering the absolute number of false positives, one can consider the percentage of false positives in the list of genes declared as differentially expressed. Methods of controlling the False Discovery Rate (FDR) are simple to implement and are available in many statistical software packages.

8.5   Maximum Likelihood

The accuracy of conclusions drawn from a statistical data analysis depends to a large degree on the quality of the model fitted to the data. The general form of the model (e.g., linear regression, phylogenetic tree, etc.) is largely dictated by the experimenter's prior knowledge and the scientific question that an experiment is designed to answer. However, the fine-tuning of a model through the appropriate choice of model parameters may be based on the outcome of the experiment (data).

Example 8.8

Whether or not it is appropriate to model the life span of fruit flies as a linear function of the thorax size has to be decided by the experimenter based on his experience and the fit of the data to the hypothesized model.

$$(\text{life span}) = \beta_0 + \beta_1 \ (\text{thorax size})$$

However, the method by which appropriate values for the intercept β_0 and the slope β_1 are found may be entirely data driven.

One of the most popular techniques for estimating the parameters of a statistical model through the observations is called the MAXIMUM LIKELIHOOD method.

GENERAL IDEA: Every statistical model will assign a probability distribution to the possible observed outcomes of an experiment that is conducted at a fixed level of predictor variables. If the values of the predictor variables change, the distribution of the response will likely change, too. Knowing this distribution allows one to compute the theoretical probability of observing any possible set of outcomes (data). The probability will depend on the parameters of the statistical model as well as the outcomes that were actually observed. Written as a function of the model parameters θ and possible observations $x_1, x_2, \ldots, x_n$, this probability is also called the likelihood function of the data

$$L(\theta; x_1, x_2, \ldots, x_n)$$

$$= \text{probability of observing } (x_1, x_2, \ldots, x_n)$$

$$\text{in a model with parameter } \theta.$$

Maximum likelihood values of model parameters may be found by maximizing the likelihood function with respect to θ. In this case, $x_1, x_2, \ldots, x_n$ are the actual observed data values.

Example 8.9

The sample proportion $\hat{p}$ is the maximum likelihood estimate of a population proportion p. Suppose we are interested in the proportion, p, of soil microorganisms that are resistant to an antibioticum. Instead of accounting for all microorganisms in the soil, we take n

soil samples, apply the antibioticum, and count the number, x, of resistant microorganisms. The sample proportion is defined as

$$\hat{p} = \frac{x}{n}.$$

If the true population proportion of microorganisms which are resistant is p, and a sample of size n represents a large population of microorganisms, then the number of resistant microorganisms in the sample has a binomial distribution (see Section 3.4.1). This means that the probability of observing x resistant microorganisms in n sample can be formulated as

$$L(p; x) = \binom{n}{x} p^x (1-p)^{n-x}$$

The former is a function of the observation x as well as the model parameter p. The number n represents the sample size and is known. To find the maximum likelihood estimate of p, we maximize the likelihood function $L(p; x)$ with respect to p (not shown here). The result is

$$\text{argmax}_p L(p; x) = \frac{x}{n}$$

where argmax refers to the argument of the maximum. That is, the maximum likelihood estimate of p is the value for which the function L attains the largest value.

8.6 Frequentist and Bayesian Statistics

As this reference Manual for bench scientists comes to a close, we address a well-known dichotomy in statistics. There are many who believe the field of statistics to be divided into two camps: the frequentists and the Bayesians. Some of us believe there are three camps, the frequentists, the Bayesians, and the "if it works, use it" group. Here we introduce the concept of Bayesian statistics in contrast to frequentist or classic statistics, which is the perspective from which this Manual is written. The basic divide between classic and Bayesian statistics lies in the definition of probability. Even though one's view and understanding of probability defines her view of classic/frequentist or Bayesian statistics, the overarching goal of statistics remains the same: Ask questions of the data based on unknown parameters.

A statistician who approaches statistics from a Bayesian perspective has a different view of probability and follows the wisdom of Reverend Thomas Bayes (1763). Bayesian statistics is rooted in thinking about uncertainty rather than long-run behaviors or probability. Statements are made with a level of certainty or belief attached to them, which in turn allows probability statements to be made about unknown parameters. Specifically, Bayes rule is a conditional statement, or posterior probability, on the parameter given the data and depends on prior information about the parameter. This prior information may be a guess or hunch; it may be theoretically justified, or it may be based on previous experience. Prior information (known as "a priori") coupled with the likelihood of the data given the unknown parameters gives rise to a level of certainty (known as a "posterior probability") that can be placed on the results.

Example 8.10

A quantitative trait locus (QTL) is a region of the genome that is associated with a quantitative trait of interest. It is desirable to locate QTL for the purpose of inferring the genes or controlling elements that may lie within them. QTL are located by taking advantage of the genetic distance between known genetic markers and statistically testing at known locations in the genome. From a frequentist approach, the null hypothesis is that there is no QTL located at a particular testing location in the genome. Data are collected, a test statistic with a known distribution is calculated, and statements about the probability of observing a result more extreme than the one provided by the data are made (i.e., p-value). Alternatively, a Bayesian approach to a QTL analysis calculates the posterior probability of a QTL being located at the testing position in the genome, based on prior information about QTL location and the likelihood of the data given the QTL. A level of certainty of a QTL being present at the testing position is provided as the result.

References

Bayes T. 1763. An essay towards solving a problem in the doctrine of chances. *Philos Trans Roy Soc* **53**: 370–418.

Craig B, Black M, Doerge RW. 2003. Gene expression data: The technology and statistical analysis. *J Agri, Biol, and Environ Stat* **8**: 1–28.

DeLongis A, Folkman S, Lazarus RS. 1988. The impact of daily stress on health and mood: Psychological and social resources as mediators. *J Pers Soc Psych* **54**: 486–495.

Guy W. 1846. On the duration of life among the English gentry. *J Stat Soc London* **9**: 37–49.

Hochberg Y, Benjamini Y. 1995. Controlling the false discovery rate: A practical and powerful approach to multiple testing. *J Roy Stat Soc Series B* **57**: 289–300.

Höfer T, Przyrembel H, Verleger S. 2004. New evidence for the theory of the Stork. *Paed Perin Epidem* **18**: 88–92.

Kerr M, Churchill G. 2001. Experimental design for gene expression microarrays. *Biostat* **2**: 183–201.

Mendel G. 1865. Experiments in plant hybridization. In *Proc Nat Hist Soc Brünn*.

Nei M, Roychoudhury AK. 1993. Evolutionary relationships of human populations on a global scale. *Mol Biol Evol* **10**: 927–943.

Newton M, Kendziorski C, Richmond C, Blattner F, Tsui K. 2001. On differential variability of expression ratios: Improving statistical inference about gene expression changes from microarray data. *J Comput Biol* **8**: 37–52.

Sies H. 1988. A new parameter for sex education. *Nature* **332**: 495.

Stewart KJ, Turner KL, Bacher AC, DeRegis JR, Sung J, Tayback M, Ouyang P. 2003. Are fitness, activity and fatness associated with health-related quality of life and mood in older persons? *J Cardiopul Rehab* **23**: 115–121.

Tukey JW. 1980. We need both exploratory and confirmatory. *Amer Stat* **34**: 23–25.

USDA National Nutrient Database for Standard Reference, release 18. 2005. Iron Content of Selected Foods per Common Measure. U.S. Department of Agriculture, Agricultural Research Service.

Vaillant GE. 1998. Natural history of male psychological health, xiv: Relationship of mood disorder vulnerability to physical health. *Am J Psych* **155**: 184–191.

von Bunge G. 1902. *Textbook of physiological and pathological chemistry*, 2nd ed. Kegan Paul Trench/Trubner, London.

Yang Y, Dudoit S, Luu P, Lin D, Peng V, Ngai J, Speed T. 2002. Normalization for cdna microarray data: A robust composite method addressing single and multiple slide systematic variation. *Nucl Acids Res* **30:** e15.

Index

Index of Worked Examples

Index of Excel Commands